T0358260

Santa Bárbara's Legacy: An Environmental History of Huancavelica, Peru

Brill's Series in the History of the Environment

Series Editor

Aleks Pluskowski

VOLUME 5

The titles published in this series are listed at *brill.com/bshe*

Santa Bárbara's Legacy:
An Environmental History of
Huancavelica, Peru

By

Nicholas A. Robins

BRILL

LEIDEN | BOSTON

Cover illustrations: Portal: entrance to the Bethelhem Adit, the main entrance to the Santa Barbara Mines (front cover), View on the Huancavelica (back cover). Photos by the author.

Library of Congress Cataloging-in-Publication Data

Names: Robins, Nicholas A., 1964- author.
Title: Santa Bárbara's legacy : an environmental history of Huancavelica,
 Peru / by Nicholas A. Robins.
Description: Leiden ; Boston : Brill, 2017. | Series: Brill's series in the
 history of the environment, ISSN 1876-6595 ; volume 5 | Includes
 bibliographical references and index.
Identifiers: LCCN 2017008349 (print) | LCCN 2017011168 (ebook) | ISBN
 9789004343795 (E-book) | ISBN 9789004339781 (hardback : acid-free paper)
Subjects: LCSH: Huancavelica (Peru)--Environmental conditions. | Mercury
 mines and mining--Environmental aspects--Peru--Huancavelica--History. |
 Mercury mines and mining--Health aspects--Peru--Huancavelica--History. |
 Indians, Treatment of--Peru--Huancavelica--History. | Environmental
 degradation--Peru--Huancavelica--History. | Hazardous
 wastes--Peru--Huancavelica--History.
Classification: LCC GE160.P4 (ebook) | LCC GE160.P4 R63 2017 (print) | DDC
 363.17/91--dc23
LC record available at https://lccn.loc.gov/2017008349

Typeface for the Latin, Greek, and Cyrillic scripts: "Brill". See and download: brill.com/brill-typeface.

ISSN 1876-6595
ISBN 978-90-04-33978-1 (hardback)
ISBN 978-90-04-34379-5 (e-book)

This book is dedicated to the residents of Huancavelica, and especially to the families who have supported and participated in the field research upon which much of this work is based.

∴

Contents

Acknowledgements

I am deeply indebted to many people who have provided invaluable support and insights as I researched and wrote this book. My interest in Huancavelica's environmental history, and the toxic legacy with which it is burdened, was stimulated by my colleague Rubén Darío Espinoza Gonzales, MA. With his generous spirit and the keen eye of an archaeologist, he first brought the issue of heavy metal contamination in Huancavelica's earthen homes to my attention, and has been involved in this research ever since. Central to the process which has unfolded has been Dr. Máximio Enrique Ecos Lima, whose commitment to public health is matched by his expertise, organizational ability, and unflagging energy.

Likewise, Bryn Thoms, RG, has played a vital and pivotal role in the field research which inform this work. Indefatigable, enthusiastic and precise, Bryn embraces, and overcomes, the challenges that field work presents. Moreover, he has taken the lead in preparing Huancavelica's remedial investigation and related reports, and painstakingly assembled the tables which form the appendix of this work. Also a key member of our field research team is Dr. William E. Brooks, whose acumen and perspective have enriched the experience in many ways. All of these colleagues have been instrumental in developing our understanding of Huancavelica's current heavy metals contamination. Not only does their knowledge inform this book, but their dedication, compassion, and curiosity bring a wonderful and collegial spirit to our work.

In the Office of Epidemiology of Huancavelica's Zacarías Correa Valdivia Hospital, I have also been immensely fortunate to be able to rely on Isidora Cauchos, Elsa Matamoros, Vilma Cauchos, Javier Quezada, Jorge Moreira Tora and Yolanda Castellares Aramburú for crucial and consistent community outreach and logistical support in the field. Their intimate knowledge of, concern for, and commitment to the people of Huancavelica is inspiring. Likewise, I would like to thank the Directors of the Hospital, Dr. Wendy Pompilio and Dr. Julio Álvarez, for their support of this research.

At the National University of Huancavelica, Dr. Nicasio Valencia Mamani, Dr. Zeida Patricia Hoces La Rosa, and Dr. Rossibel Muñoz de la Torre have offered unceasing support for this research, and have contributed greatly to a wider public understanding of legacy contamination issues in Huancavelica. Similarly, Dr. Pedro de la Cruz, Dr. Jesús Mery Arias Huánuco, Dr. Ruggerths de la Cruz and Dr. Omar Siguas have played important roles in broadening awareness of environmetal issues in the city. Also in Peru, I woud like to thank Dr. Paulo Vilca and Barbara Fraser, who have so often provided welcome advice and insights as I prepared this work.

I would also like to express my gratitude to Dr. Robert Robins for comments on an earlier draft of this book, and to my colleagues at North Carolina State University (NCSU), Dr. David Zonderman, Dr. Steven Vincent, Dr. Matthew Booker and Dr. Richard Slatta for their interest in, and support of, my research. I am also grateful for the support of Ms. Sydney Thompson and the exceptional staff of the Interlibrary Loan Department at the D.H. Hill Library at NCSU for their diligent assistance in obtaining obscure publications. Similarly, I am indebted to Mr. Edwin Diáz, Director of Research and Bibliographic Services at the National Library of Peru, who kindly facilitated my research there.

Earlier research related to this volume was enabled by an Oak Ridge Institute for Science and Education Fellowship at the Environmental Protection Agency, for which I am especially indebted to Dr. John Vandenberg. Similarly, Dr. Nicole Hagan, Dr. Heileen Hsu Kim, Dr. George Woodall and Dr. Daniel Richter have all played important roles in the research upon which this work is based, for which I am grateful. Similarly, I am thankful for the support provided by Kendall Brown, Judith Teran Ríos, Rubén Julio Ruíz Ortíz, Joaquín Loayza Valda, Maria del Carmen Martínez López, Alvaro López Donoso, Leonor Ferrufino Fernández, Oscar Hurtado Borja, Cecilia Mardoñez Barrero, Ana María Nava, María Renee Pareja Vilar, José Antonio Fuertes López, Silvia Flores, Luis Tórrez Ameller and Sheila Beltrán Lopéz.

I would also like to thank Aleks Pluskowski, Wendel Scholma and Anita Opdam at Brill Publishers for the opportunity to work with them as this project has progressed through the review and publicaton process.

I am especially grateful for the unceasing support for, and encouragement of, this project provided by my wife and collaborator, Dr. Susan Halabi.

Most of all, I would like to thank the individuals who have opened their earthen homes to us as part of our field work, and patiently allowed us to chip away samples of their walls and floors for analysis. Their kindness is exemplary, and their suffering unjust.

While I am grateful to all, any errors or omissions in this work remain my own.

Glossary

Alcalde	A mayor of a municipality who had judicial powers.
Alcalde mayor de minas	A magistrate with jurisdiction over mining issues.
Alguacil	A sheriff.
Allyu	An extended kin group descended from an often fictional ancestor.
Alquila	A wage laborer.
Apu	A male, Andean mountain spirit.
Arancel	A schedule of fees for religious services or reparto goods.
Asiento	A multi-year agreement between the mining guild of Huancavelica and the crown to produce a fixed amount of mercury for which the crown would pay a set price.
Audiencia	A regional court with executive and limited legislative powers.
Barretero	A wage earning mine worker who extracted ore from the mine face.
Bola	A ball made of mercury-containing ash and solid discharge from various tubes in a mercury smelting operation, which was then re-smelted.
Buitrón	The walled "patio" in which torta was deposited and where amalgamation took place.
Buscón	A prospector.
Busconil	A smelter developed by Lope de Saavedra Barba, and later used in Almadén where they became known as Bustamante smelters.
Cabildo	A city council.
Cajón	A tub in which the silver bearing ore, mercury, water and other ingredients were mixed prior to being spread out in the amalgamation process. In Huancavelica it referred to a load of cinnabar of approximately sixty-eight kilograms.
Capitán enterador de mita	A "mita capitain" who accompanied mitayos to their destination and had the fiscal responsibility to see that they completed their terms.
Capitán general de mita	An Indian who oversaw the work of the capitanes enteradores de mita in a province.

Carguiche	A worker who carried ore to the surface.
Chaccu	An annual roundup of vicunas for shearing of their wool.
Chacanea	A worker who ran llama trains to carry ore from the pithead to smelters.
Chakitaklla	A wooden device with a curved point and cross member upon which the foot is placed, utilized for planting and harvesting tubers in the Andes.
Chicha	A fermented beverage made from corn.
Chunca Horno	The place of ten ovens, from Quechua, chunca, which is ten, and Spanish, horno, oven. This was the San Cristóbal neighborhood in Huancavelica where mitayos lived and much smelting took place.
Chuño	A freeze dried potato.
Cofradía	A religious brotherhood organized in worship of a specific saint.
Composición de tierras	A process of land titling which resulted in Indian community lands being owned by non-Indians.
Conopas	Huacas associated with fertility.
Corregidor	A governor of a colonial district.
Curaca	The leading authority of an indigenous community.
Cuy	*Cavia porcellus*, a guinea pig.
Encomendero	A holder of an encomienda.
Encomienda	A grant of Indian labor to a Spaniard, usually a conquistador, in exchange for which the beneficiary agrees to Christianize his charges and defend the realm when necessary.
Endiabladas	The practice of opening and cleaning a smelter before it had sufficiently cooled, exposing the worker to mercury vapor.
Estribo	Literally, a stirrup, or a stone column inside a mine used for structural support.
Faena	An often unpaid task assigned to an Indian, often on urban public works or in mines and involving upkeep, improvements or cleaning.
Forastero	An Indian who did not live in his town of birth.
Gremio	A guild.
Guasacho	A piece of cinnabar ore, informally considered part of a wage, which was then sold or exchanged by mine

	workers for chicha or other goods or money at the mine entrance.
Guayra	A small ceramic or stone wind-blown furnace utilized for refining rich silver ore in the Andes.
Hacendado	An owner of an hacienda.
Hacienda	An agricultural estate.
Hilacata	A council of leaders of an indigenous community.
Huaca	An often local deity or shrine to the same.
Huallata	*Chloephaga melanoptera*, or Andean goose.
Humanchi	An independent, small scale miner, also known as a pallequeador.
Indio de faltriquera	A practice whereby a mitayo Indian pays a sum of money directly to the individual to whom he is ordered to work to be released from service.
Indio de plaza	A fixed-term forced laborer dedicated to non-mining tasks.
Indio en plata	Term used to describe an the practice of sending money instead of a mitayo to the beneficiary of a mitayo allotment.
Intendant	A governor, based on the French intendant system, which replaced that of corregidors in the late eighteenth century.
Inti	The Andean sun god.
Jabeca smelter	A smelter consisting of a fire chamber which stood around a meter high, and which had openings on the top which received the lower part of the refining vessels, the lids of which led to condensation tubes.
Ladino	An indigenous person who has adopted Hispanic cultural characteristics.
Lamas	The poorest grade impurity laden mercury-silver amalgam contained in runoff from the washing phase of the amalgamation process.
Lavadero	A place on a streambank or near a spring where pre-Hispanic groups processed cinnabar to produce vermilion.
Los rosa	Indian laborers assigned to mine maintenance tasks.
Llimpi	Vermilion extracted from pulverized cinnabar which was utilized as cosmetic, paint and ritualistic purposes in the Andes prior to the Spanish conquest.

Mama Quilla	The Andean moon goddess.
Maray	A mortar, often made of granite, with the lower, concave, piece being around one meter in diameter and the upper, grinding, stone about one-half that size.
Mestizo	A person of Indian and Spanish descent.
Minería del Rey	The Royal Mining Company which directly operated mines in Huancavelica and was organized by Antonio de Ulloa in the mid-1700s.
Mita	Compulsory Indian labor system, utilized for mining in Huancavelica and Potosí. Literally a turn of service in Quechua.
Mitayo	Indian laborer serving a mita.
Muqui	A masculine Andean deity who lives underground and determines the fate of miners.
Obraje	A vertically integrated textile production center.
Oidor	A judge on an audiencia.
Originario	An Indian who lives in his birth-town.
Oyarico	A cinnabar smelter operator.
Pachacuti	According to Andean belief, the inevitable, cyclical destruction of the world.
Pachamama	Mother Earth, the earth deity.
Palleaquedor	An individual engaged in pallequeo, also known as a humanchi.
Pallequeo	In Peru, freelance or informal mining.
Piña	A mold into which mercury-silver amalgam was placed for firing.
Piquero	A pick-man.
Pella	Mercury-silver amalgam.
Potabambas	Another term for los rosa, or Indian laborers assigned to mine maintenance tasks.
Protector de los Naturales	Legal advocate for the Indians
Porina	A wage to cover the time it took for mitayos to come to and from Huancavelica.
Puente	Literally, a bridge, or a structural stone archway in a mine.
Punchao	A load of cinnabar ore which measured about one and one half metes high and a half meter feet wide, in theory a day's work for a carguiche.
Qhapaq Ñan	Inca road linking Cuzco to Quito.

Quimbalete	The grinding, upper, stone element used with a maray to pulverize cinnabar.
Quintal	A weight equivalent to 101.19 pounds or 45.89 kilograms
Quinto Real	Literally the "royal fifth," a tax collected on mineral production. The actual percent charged varied by location and time, and was occasionally reduced to stimulate production.
Real	One-eighth of a peso of eight.
Reducción	A settlement patterned after the Spanish urban layout with a central plaza into which Indians were forcibly relocated after the conquest.
Ribera	The canal which channeled water from the lagoon reservoirs above Potosí to the chain of silver refining mills throughout the city.
Regidor	A city councilman.
Relave	The richest mercury-silver amalgam in runoff from the washing phase of the amalgamation process.
Relavillo	Intermediate grade mercury-silver amalgam in runoff from the washing phase of the amalgamation process.
Repartimiento de mercancías	The forced purchase of goods by Indians operated by corregidores. Also known as the reparto.
Reparto	Shortened term referring to the repartimiento de mercancías.
Repasari	Indian workers who would tread the torta to facilitate amalgamation.
Sirvancuy	Andean trial marriage.
Sobrestante	A veedor's assistant, charged with ensuring the safe and orderly operation of a mine.
Soldados	Technically a soldier, however as used in a mining context it referred to a mobile male population known for being unruly and seeking quick fortunes.
Tambo	An inn which indigenous communities were required to staff and maintain.
Taquia	Llama excrement, used as fuel.
Taruca	*Hippocamelus antisensis*, a deer native to the Andean region.
Torta	The mixture of silver bearing ore, mercury, water salt and other elements in which the amalgamation process takes place.

Tren Macho	Nickname of the train linking Huancavelica to Huancayo.
Umpé	Pockets of carbon monoxide or carbonic acid inside a mine.
Veedor	An inspector with the responsibility of ensuring the safe and orderly operation of a mine.
Villa	A municipal district, literally a town.
Visita	A working tour of a jurisdiction by a royal official, such as a viceroy.
Visitador	A representative of the king with extensive powers, often in a specific area.
Vizcacha	*Lagidium peruanum*, a long tailed rodent resembling a rabbit and related to the chinchilla.
Wanka	A stone with religious significance.
Yanacona	An Indian servant of Spanish conquerors, later associated with serfdom on agricultural estates.
Yanavico	*Plegadus ridgwayi*, or Puna Ibis.
Yerba maté	*Ilex paraguariensis*, an evergreen with caffeinated leaves which are consumed like a tea.

Introduction

One afternoon in 1850, Indians digging a ditch in the town of Huancavelica, Peru, witnessed an impressive event. Their parish priest had directed them to excavate a channel to divert rainwater away from the church. Hardly had they penetrated a meter below the surface when they noticed mercury droplets seeping from the sides of the trench. The quicksilver would continue to trickle over the next two weeks, during which time workers collected almost two metric tons of the valuable metal. Although it was not the first time that native, or naturally occurring, mercury had been encountered in the town, it underscored the ongoing, intimate and toxic links between the city's residents and their natural environment.[1] The effects of that relationship are not, however, limited to the town, and have had far-reaching ecological, economic and public health consequences.

Nestled in a sharp valley incised high in the Peruvian Andes, Huancavelica straddles one of the world's largest cinnabar deposits. Rivaling those of Almadén in Spain, Monte Amiata in Italy and Idrija in Slovenia, the mines of Huancavelica have rendered over sixty thousand metric tons of mercury, more than any such deposit in the western hemisphere.[2] Huancavelica's cinnabar deposits served as a source of colorants for pre-Hispanic societies which used vermilion for ritualistic and other purposes. During the Spanish colonial era, indigenous laborers were conscripted to extract cinnabar and refine it for mercury, the ineluctable ingredient which fueled the region's amalgamation-based silver mills. The mercury amalgamation system enabled silver refiners to process lesser quality ores, which although abundant in the Andes, were unsuitable for smelting. Just as mercury amalgamates with silver, so too has mercury amalgamated the Huancavelica's environmental history with global economic history. The industrial scale silver production which mercury enabled led to an unprecedented and centuries-long flow of silver to Europe, the Middle East and the Orient. This current of silver, in turn, spurred the establishment the

---

1   León Crosnier, "Geologie du Perou. Notice geologique sur les Departments de Huancavelica et d'Ayachcho," in *Annales de mines*. 5th series, Vol. 11 (1852), 2, 51–52; Edward Berry and Joseph T. Singewald, *The Geology and Paleontology of the Huancavelica Mercury District* (Baltimore: Johns Hopkins Press, 1922), 43.

2   Patricia D'Itri and Frank D'Itri, *Mercury Contamination: A Human Tragedy* (New York: John Wiley and Sons, 1977), 118; E.H. McKee, D.C. Noble and Cesar Vidal. "Timing of Volcanic and Hydrothermal Activity, Huancavelica Mercury District, Peru," in *Economic Geology and the Bulletin of the Society of Economic Geologists*. Vol. VIII1, No. 2 (1986), 489.

first global trade networks, and ultimately the Industrial Revolution and to-day's globalized economy.

Beyond global economic integration, mercury refined in Huancavelica has bequeathed an enduring toxic legacy in the soils and waterways of the city, earn-ing it the distinction of being one of the world's most mercury-contaminated urban areas. Lead and arsenic also lace the city's soils, and, along with mer-cury, are found in the adobe or rammed earth homes in which about half of the city's population currently live. Among the results is an estimated 19,000 people living in homes which have mercury contaminated soils or vapor above World Health Organization screening levels, or levels that are determined to be safe for lifetime residential exposure.[3] Today, Huancavelica is Peru's poor-est departmental, or state, capital, saddled with the nation's highest rates of extreme poverty, female illiteracy, and malnutrition among children under five years old.[4]

Based on archival and printed primary documents, secondary sources and scientific research, this book traces the origins of this toxic legacy, and the mu-tually conditioning relationships between the environment and the human population of the city and region of Huancavelica, from pre-history to the pres-ent. While such interactions are inherently local, in the case of Huancavelica, their effects were both regional and global. Public policy decisions made in

---

3   Nicholas Robins, N. Hagan, S. Halabi, H. Hsu-Kim, R.D. Espinoza Gonzales, M. Morris, G. Woodall, D. Richter, P. Heine, T. Zhang, A. Bacon, and J. Vandenberg, "Estimations of His-torical Atmospheric Mercury Concentrations from Mercury Refining and Present-Day Soil Concentrations of Total Mercury in Huancavelica Peru," in *Science of the Total Environment*. Vol. IV26, No. 11 (June, 2012), 152; Nicole Hagan, Nicholas Robins, Heileen Hsu-Kim, Susan Halabi, Rubén Darío Espinoza Gonzales, Daniel Richter and John Vandenberg, "Residential Mercury Contamination in Adobe Brick Homes in Huancavelica, Peru," in *PLoS ONE*, Vol. VIII No. 9. (September, 2013), 1; Bryn Thoms, Nicholas A. Robins, Enrique Ecos, Earl W. Brooks and Rubén Darío Espinoza Gonzales. *Results of June–July, 2015 Field Study, Huancavelica, Peru*, (Environmental Health Council, Unpublished manuscript. 2015), 6, 11, 12; Bryn Thoms, Nicho-las Robins, Enrique Ecos, and William E. Brooks. *Results of June/July 2016 Assessment of Soil and Fish, Huancavelica Mercury Remediation Project*, (Environmental Health Council, unpub-lished manuscript, November, 2016), 9–10. See also Tables 1, 2, 3, 5, 6 and 8 in appendix.

4   Government of Peru, *Censos Nacionales 2007, XI De Población Y VI De Vivienda: Resulta-dos Definitivos*. (Lima: Instituto Nacional de Estadística e Informática, 2008), 941; Govern-ment of Peru, *Estado de la Población Peruana, 2014* (Lima: Instituto Nacional de Estadística e Informática, 2014), 7; Government of Peru, Instituto Nacional de Estadística, Sistema de Consulta de Principales Indicadores de Pobreza, Mapa de Pobreza, http://censos.inei.gob .pe/Censos2007/Pobreza/; Government of Peru, *Encuesta Demográfica Y De Salud Familiar 2000*: Huancavelica. Vol. VIII, 180–181, 195–198, 203; Tesania Velázquez, *Salud Mental En El Perú: Dolor Y Propuesta: La Experiencia de Huancavelica* (Lima: Consorcio de Investigación Económica y Social, 2007), 68.

Spain and Lima reflected imperial imperatives concerning Andean silver production, and had drastic effects on the region's residents and environment. Unfortunately, in Huancavelica, this has historically been based on the production and outflow of a toxic mineral resource which entailed, and continues to extract, extraordinary human, social and environmental costs.

Within this context, the following pages examine the geological genesis of the region, its spiritual significance to pre-Hispanic societies and their broader relation to the environment, and how the Spanish conquest affected the biological, ecological, demographic and economic contexts of society. In so doing, the volume details the rise of the mercury amalgamation system which allowed sustained, industrial scale silver production, and the effects this had on indigenous communities, the environment, public health, labor relations and the global economy. The work also explores the environmental and social history of the region in the nineteenth century, a period of limited mercury production, ecological recovery, and regional isolation. Twentieth century efforts to revive production, and the enduring and multifaceted legacy of heavy metals contamination on the environment and people round out the study.

The Nature and Uses of Mercury

Mercury is primarily contained in reddish cinnabar ore, or mercuric sulfide, and is as old as the earth itself. It has been found on all of the planet's major landmasses and is associated with igneous, or volcanic, rock, or that of volcanic origin, as well as with thermal springs. Cinnabar is commonly found in sandstone and limestone, and must be retorted at 580 degrees Celsius for it to yield mercury vapor, before condensing into liquid form.[5] Although quicksilver is also released into the atmosphere as a result of the natural erosion of mercury-containing rocks, anthropogenic emissions, or those caused by humans, account for most of that which has entered the atmosphere. Sources include forest fires, which volatize mercury in soils, electricity plants which use mercury-containing coal, and the production of mercury for industrial

---

5   Crosnier, 37–38, 44; D'Itri, 118, 120; Hugh Evans, "Mercury," in *Environmental and Occupational Medicine*. 3rd. Ed. William Rom, Ed. (New York: Lippincott- Raven Publishers, 1998), 997; Leonard J. Goldwater, *Mercury: A History of Quicksilver* (Baltimore: York Press, 1972), 2–4, 32, 49; United States Environmental Protection Agency, *Mercury Study Report to Congress*, Vol. I. Executive Summary (Washington, D.C.: United States Environmental Protection Agency, 1997), 2–1; Nicholas Robins, *Mercury, Mining and Empire: The Human and Ecological Cost of Colonial Silver Mining in the* Andes (Bloomington: Indiana Univerity Press, 2011), 101; William E. Brooks, Estevan Sandoval, Miguel Yépez and Howell Howard, *Peru Mercury Inventory, 2006*, (Washington, DC: United States Geological Service, 2007), 4.

purposes. Currently, about 5,500 metric tons of mercury are released annually by both natural and human actions, and about ten to twenty percent of this is as a result of the artisanal gold production.[6]

Once volatized, mercury may remain aloft in the earth's atmosphere for months, either as a gas or bound to particles. Inevitably, it will descend to earth, usually with rainfall. While much of this will alight in oceans, some will deposit on landmasses and adhere to soils. A portion of this mercury may revolatize, while another part may be swept away by rainwater and settle in rivers and lakes. There it may be consumed by anaerobic microbes, which convert it into ultra-toxic methylmercury. Unless it is encapsulated, once in the water mercury may be consumed by mollusks and fish and bio accumulate in the aquatic food chain.[7]

The Inca and their forebears were not the only ones to use cinnabar for cosmetic and ritualistic purposes, as well as a colorant for dyes and paint. The practice is found in many cultures and goes back for centuries. For example, archaeologists uncovered a 3,500 year old sample of cinnabar in Egypt, and the red ore was obtained by Phoenician traders 2,700 years ago. Some of the earliest evidence of retorting mercury was found in Greece, where it was used for medical and ritualistic purposes around 300 BC. It appears, however, that the Romans were the first Europeans to discover its usefulness in refining silver and gold, and in gilding and making mirrors. After 201 BC, Roman control of the Almadén cinnabar deposit in present-day Spain ensured a plentiful supply of quicksilver for the empire. It was also used for medical and colorant purposes in China, and by 1000 AD it was used in the Arab world to treat skin conditions. By the early 1500s, European physicians had found it was useful for treating syphilis, a practice which would continue until the early twentieth century.[8] As the uses of mercury increased, so too the recognition of its toxicity. In the European world, by the 1470s physicians understood that mercury vapor was harmful, and identified uncontrolled shaking, muscle cramping, hyper-salivation, gingivitis and tooth loss as symptoms of mercury poisoning.[9]

---

6   L.D. Lacerda, and R.V. Marins, "Anthropogenic mercury emissions to the atmosphere in Brazil: The impact of gold mining," in *Journal of Geochemical Exploration*, Vol. V8 (1997), 224; USEPA, *Mercury Study*, o–1; D'Itri, 118; Hugh Evans, "Mercury," 997; Robins, *Mercury, Mining and Empire*, 102.

7   USEPA, *Mercury Study*, o–1, 2–1, 2–4, 3–1; Robins, *Mercury, Mining and Empire*, 102.

8   John Frith, "Syphilis: Its Early History and Treatment Until Penicillin, and the Debate on its Origins," in *Journal of Military and veterans Health*. Vol. IIo, No. 4 (November, 2012), 52–53; D'itri, 6–7; Robins, *Mercury, Mining and Empire*, 102.

9   D'Itri, 5–7, 106, 120, 122; Lesley P. Bidstrup, *Toxicology and Mercury and its Compounds* (New York: Elsevier Publishing Company, 1964), 1; na. "Health Effects," in *Toxicological Profile for*

Despite its toxic properties, mercury has found many modern uses in medicine and industry, and continues to be used in some religious rituals, such as Santería. Quicksilver is especially sensitive to, and useful in measuring, atmospheric pressure and temperature, hence its use in barometers and thermometers. In the nineteenth century, metallic mercury was an ingredient in fingerprint powders, and mercury nitrate was employed in the treatment of rabbit pelts to make felt hats, practices which poisoned both police investigators and hat makers alike. More recently, mercury has been, and in some places still is, used in cosmetics such as skin whitening agents. Mercury halogen bulbs contain mercury, as do florescent lights, and it has also been employed in the production of batteries, chlorine, paint, paper, medical devices, amalgams for dental fillings, and explosive detonators. In agriculture, it has found applications to control insects and fungi. In the early twenty-first century increasing gold prices spurred demand for mercury in artisanal gold mining operations. Despite this, mercury is being phased out of industry given its deleterious effects on humans and the environment.[10]

The Broader Context: The Global Effects of Andean Silver Production

Although quicksilver has numerous uses, among its most prolific applications has been in silver production. Much of the silver exported from South America during the colonial and early national periods would not have been released from its ore were it not for Huancavelica's mercury. Just as quicksilver vapor

---

Mercury. Agency for Toxic Substances and Disease Registry, United States Department of Health and Human Resources (Atlanta, GA: United States Department of Health and Human Resources, 1999), 226; L. D. Lacerda, "Global mercury emissions from gold and silver mining," in *Water, Air and Soil Pollution* Vol. IX7 (1997), 210; Goldwater, *Mercury*, 24, 269–270; Robins, *Mercury, Mining and Empire*, 103; Alfredo Menéndez Navarro, *Un mundo sin sol. La salud de los trabajadores de las minas de Almadén, 1750–1900* (Granada: Universidad de Granada, 1996), 78–79.

10 D'Itri, 3, 55, 105–108, 136; Marsh, 80; Eduardo de Habich, "Industria de azogue," in *Boletín de Minas, Industrias y Construcciones.* Vol. II (Lima, 1885), 12; Goldwater, *Mercury*, 127–128, 270; Norbert Schutte, et al., "Mercury and its Compounds," in *Occupational Medicine* 3rd. Ed., Carl Zenz, et al. Eds. (St. Louis, MO: Mosby, 1994), 550–551; H. A. Waldron and A. Scott, "Metals," in *Hunter's Diseases of Occupations*, 8th Ed. P.A.B. Raffle et al., Eds. (London: E. Arnold, 1994), 102; "Health Effects," 125; USEPA, *Mercury Study*, 2–1, 3–7; USEPA, *Mercury Study*, 0–3, 3–7, 3–8; Lacerda, "Global mercury emissions," 210; Robins, *Mercury, Mining and Empire*, 103.

from Huancavelica flowed in the earth's atmosphere, invisibly settling and contaminating countless places, the Spanish silver pesos of eight which the amalgamation system yielded circled the globe and expanded economic frontiers. The first economic impact of the mercury-silver amalgamation system was, however, regional, as surging silver production in the Andes spawned innumerable economic linkages. This was most notable in the region of Potosí, in present-day Bolivia, which in the last quarter of the sixteenth century produced about one-half of all silver in the Americas and almost three-quarters of that of the Andes.[11]

Like Potosí, most Andean silver mining towns were located on the altiplano, where few crops are cultivated beyond tubers and quinoa. The dearth of other resources in such mining centers created a strong demand for a plethora of products ranging from construction and mining inputs to foodstuffs, clothing and imported luxury items. Beyond the demand for such materials, and their consequently high cost, silver mining centers were especially attractive markets as they had ready money. This combination spawned the growth of *haciendas*, or agricultural estates, and *obrajes*, or vertically integrated textile enterprises, which, along with vineyards and ranches, produced countless items for highland mining towns such as Potosí, Oruro and Castrovirreyna. The value of colonial American mining markets paled, however, in comparison to that of their export: over 86,000 metric tons of silver from the Americas during the colonial period ending in 1810. This silver, both legal and contraband, lumbered across the Atlantic on ships, ultimately destined not only to Spain, but also France, England, Italy, Russia and other countries.[12]

---

11   David A. Brading and Harry E. Cross. "Colonial Silver Mining: Mexico and Peru," in *The Hispanic American Historical Review*, Vol. V2, No. 4 (November, 1972), 571; C. Sempat Assadourian, "La crisis demográfica del siglo XVI y la transición del Tawantinsuyo al sistema mercantil colonial," in *Población y mano de obra en América Latina*, Nicolás Sánchez Albornoz, Ed. (Madrid: Alianza Editorial, 1985) 69; Carlos Marichal, "The Spanish-American Silver Peso: Export Commodity and Global Money of the Ancien Regime, 1550–1800," in *From Silver to Cocaine: Latin American Commodity Chains and the Building of the World Economy, 1500–2000*, Steven Topik, Carlos Marichal and Zephyr Frank, Eds. (Durham, NC: Duke University Press, 2006), 25.

12   John J. TePaske and Kendall Brown, Kendall W., Eds. *Atlantic World, Volume 21: A New World of Gold and Silver* (Boston, MA, USA: Brill Academic Publishers, 2010), 16, 305; Carlos Marichal, "The Spanish-American Silver Peso: Export Commodity and Global Money of the Ancien Regime, 1550–1800," in *From Silver to Cocaine: Latin American Commodity Chains and the Building of the World Economy, 1500–2000*. Steven Topik, Carlos Marichal and Zephyr Frank, Eds. (Durham, NC: Duke University Press, 2006), 29; Dennis Flynn and Arturo Giraldez, "China and the Manila Galleons," in *World Silver and Monetary History in*

While the Spanish crown sought to impose a trade monopoly upon its colonies, its ability to do so was attenuated by the paucity of goods produced in Spain, and the attractive market which the Americas offered to European rivals. During much of the colonial era, Spain produced little else than wine, olives, iron and wool. Efforts to prohibit the production of grapes and olives in South America were widely ignored, as olive orchards and vineyards sprung up in the region, especially on Peru's coastal plain. Although silver producers relied on iron, usually from Vizcaya, to minimize mercury consumption in refining, people had little use for Spanish wool given the high quality fibers of the abundant alpaca and vicuña population in the Andes.[13]

The result was that Spain essentially became an intermediary between the Americas and Europe, as foreign merchants worked through Spanish brokers to legally export products to the colonies. Consequently, much of the profit from the sale of these goods did not remain in Spain, but were remitted elsewhere. Despite this, the surge of silver from the Americas resulting from the introduction of the amalgamation system provided the crown with unprecedented revenue in the form of the royal tax on minerals production as well as a host of other levies and fees. For example, in 1570, prior to the use of amalgamation in Potosí, the treasury office there collected only 11,000 pesos as the *quinto real*, or royal fifth. Subsequent to the introduction of the system, in 1579, receipts had increased over 900%, to over 1,000,000 pesos.[14]

This outpouring of silver from the Americas had several effects. One was a view of the Americas by the Spanish monarchy as an inexhaustible font of wealth, for even as some mines declined in production, there was always the expectation that imminent strikes would reveal new riches. This assumption discouraged investment in productive enterprises in Spain, and led to a dangerous dependence upon a mineral monoculture. In addition, this belief fueled imperial ambitions and the Counter Reformation, leading the Crown to borrow excessively in Europe and saddling it with debt instead of economic

the 16th and 17th Centuries (Brookfield, VT: Variorum, 1996), 86; Earl Hamilton, _American Treasure and the Price Revolution in Spain, 1501–1650_ (Cambridge: Harvard University Press, 1934), 71; Robins, _Mercury, Mining and Empire_, 4.

13   Gwendolyn Cobb, _Potosí y Huancavelica, bases económicas, 1545–1640_ (La Paz: Banco Minero de Bolivia, 1977), 94, Attman, 35; Stanley and Barbara Stein, _Silver, Trade and War: Spain and America in the Making of Early Modern Europe_ (Baltimore: Johns Hopkins University Press, 2000), 34.

14   Stein, 15, 23, 36, 52, 77–86, 261, 265; Attman, 7–8, 23, 30, 35, 39, 53, 58, 60–61; Hamilton, 305; Marichal, 38; Arthur Zimmerman, _Francisco de Toledo, Fifth Viceroy of Peru, 1569–81_ (Caldwell,Idaho: The Caxton Printers, Ltd., 1938), 132; Robins, _Mercury, Mining and Empire_, 5, 39.

development. Another effect was price inflation, not only in Spain, but in Europe more generally as more specie entered the market.[15]

While much American silver passed through Spain on its way to a foreign creditor, a considerable amount bypassed the Iberian Peninsula altogether as a result of illicit trade with European merchant vessels on Latin America's coasts, direct trade with Asia, and piracy.[16] Despite Spain's efforts to control trade through monopolies and periodic shipments of merchandise through the fleet system, the provision of goods was highly inconsistent and could never meet demand. Contraband filled the gap, as French, British and Dutch ship captains colluded with local officials eager to supplement their meager salaries with bribes and merchandise. Contraband was both ritualistic and rhythmic, as ships entered American ports under ostensible duress, captains negotiated with local officials, and merchandise was sold. It was also rhythmic, as captains would often time their arrival off of the Chilean or Peruvian coasts at peak times of silver production.[17]

American silver traveled the world, often in the form of the Spanish peso of eight, called such because it was equivalent to eight *reales*, a smaller denomination. Pesos of eight were often debased and poorly minted, however

15   G.N. Clark. "The Early Modern Period," in *The European Inheritance*. Vol. 11, Earnest Barker et al., Eds. (London: Clarendon Press, 1954), 79; Dennis Flynn, "A New Perspective on the Spanish Price Revolution: The Monetary Approach to the Balance of Payments," in *World Silver and Monetary History in the 16th and 17th Centuries* (Brookfield, VT: Variorum, 1996), 389; Iden, "Fiscal Crisis and the Decline of Spain (Castile)," in *World Silver and Monetary History in the 16th and 17th Centuries* (Brookfield, VT: Variorum, 1996), 147; Stein, viii, 3–4, 8, 27, 41, 53–54, 104–105, 144, 264–265; Attman, 30, 33, 68; Carlo M. Cipolla, *Conquistadores, piratas, mercaderes: La saga de la plata española*. Ricardo González, Trans. (Buenos Aires: Fondo de Cultura Económica de Argentina, 1999), 53; Hamilton, 44, 72, 192–211, 261, 281, 301–302, 305; Marichal, 37; Robins, *Mercury, Mining and Empire*, 5.

16   Stein, 27, 36; Marichal, 38, 41–42; Cipolla, 63–64; Flynn and Giraldez, 72, 85; Stein, 24; Gwendolyn Cobb, "Supply and Transportation for the Potosi Mines, 1545–1640," in *Hispanic American Historical Review*, Vol. II9, No. 1. (1949); 28; Cross, 412–413; Robins, *Mercury, Mining and Empire*, 5–6.

17   Arzans, Vol. III, 18, 29, 55; Audiencia de Lima, "Relación que la Real Audiencia de Lima hace al excelentísimso Sr. Marqués de Castel-Dosrius, Virey de estos reinos, del estado de ellos, y tiempo que ha gobernado en vacante," in *Relaciones de los Virreyes y Audiencias que han Gobernado el Perú*. Sebastian Lorente, Ed., Vol. II (Madrid: Imprenta y Estereotipia de M. Rivadeneyra, 1871), 293; Marichal, 38; Cipolla, 23–24; Enrique Tandeter, *Coercion and Market: Silver Mining in Colonial Potosí, 1692–1826* (Albuquerque, University of New Mexico Press, 1993), 9; 12, 96, 117, Adrian Pearce, "Huancavelica 1700–1759: Administrative Reform of the Mercury Industry in Early Bourbon Peru," in *Hispanic American Historical Review*. Vol. VII9, No. 4 (Nov. 1999), 671.

their near omnipresence meant that they were widely used nonetheless. As American silver was remitted to England, France, the Netherlands and other European countries, some of it was then invested domestically. In England, for example, New World silver allowed the capital accumulation which would enable the Industrial Revolution. What was not invested domestically was exchanged internationally as both Holland and England expanded trade routes in the Middle East, India and Asia.[18]

Most of the silver that was destined for international trade ended up in China, where it was exchanged for silks, spices, gems, tea and other luxury items. The abolition of paper currency in China in the mid-fifteenth century, and the subsequent imposition of a tax payable only in silver, increased the demand for specie there, and along with it exports of Chinese goods. The result was that by the early seventeenth century, upwards of 150 tons of silver were entering China annually. Although it was illegal, a significant percentage was dispatched from the Peruvian and Mexican coasts, where it was traded in Manila and re-exported to China. Some estimates suggest that at least one-third of the silver refined in the Americas made its way to China without ever touching the European continent. The indispensable fuel for this process of incipient economic globalization was mercury, much of which was refined in Huancavelica[19]

While mercury was vital to Spain's imperial economy, whether it was used among pre-Hispanic Andean groups raises an intriguing question. There clearly was extensive use of *llimpi*, or vermilion made from pulverized cinnabar, in pre-conquest times. This was used in paints, dyes and ornaments, as well as in burials and religious rituals. It also served as a cosmetic for elite women who would adorn their face, hands and feet with it, and also for warriors who used it as a face paint to instill fear in their enemies.[20]

---

18   Marichal, 25–26, 35; Attman, 101; Cipolla, 57, 69–72; Hamilton, 51, 53; Flynn and Giral-
      dez, 86; Hamilton, 46; Harry E. Cross, "South American Bullion Production and Export
      1550–1750," in *Precious Metals in the Later Medieval and Early Modern Worlds*. J.F. Richards,
      Ed. (Durham, NC: Carolina Academic Press, 1983), 397, 404; Cipolla, 66; Attman, 5–6, 9,
      67, 77; Marichal, 27, 38, 40–41; Stein, 26, 261–262; Cross, 397, 404; Assadorian, "La crisis
      demográfica," 69; Robins, *Mercury, Mining and Empire*, 4–6.

19   Marichal, 41–42; Cipolla, 63–64; Flynn and Giraldez, 71–72, 85; Stein, 24; Cobb, "Supply
      and Transportation," 28; Cross, 412–413; Earl Hamilton, *American Treasure and the Price
      Revolution in Spain, 1501–1650* (Cambridge: Harvard University Press, 1934), 46; Robins,
      Mercury, Mining and Empire, 6.

20   N.A., "Memoria sobre la mina de azogue de Huancavelica," in *Colección de memorias
      científicas, agrícolas é industriales publicadas en distintas épocas*. Vol. II. Mariano Eduardo
      de Rivero y Ustáriz, Ed.(Brussels: Imprenta de H. Goemaere, 1857), 85; Bernabé Cobo,

Pre-Hispanic exploitation of cinnabar from Huancavelica is substantiated by the documentary record, as early colonial miners in Huancavelica referred to the *lavaderos*, or sites on stream banks and springs where cinnabar was processed to produce vermilion. So abundant were these quicksilver-rich tailings in the immediate region of Huancavelica that in the mid-sixteenth century colonial officials promoted their further refinement by subjecting the mercury so produced to a lower tax than that produced with ore extracted from mine-shafts.[21] Early chroniclers of Huancavelica did not, however, mention the existence or ruins of retorts to refine mercury, nor has archaeological research uncovered definitive evidence of them. Despite this archaeological lacunae, pre-Hispanic groups may have encountered and collected native mercury, obviating the need for refining.

Complicating the picture is geological research which indicates that the region's atmosphere was contaminated by mercury in pre-Hispanic times. Although this could be the result of mercury released as a result of the smelting

Historia del Nuevo Mundo. Francisco Mateos, S.J., Ed. (Madrid: Ediciones Atlas, 1956), 150; Cantos de Andrade, 304; George de Fonseca, "Informe de George de Fonseca del 24 de Julio de 1622," in *Huancavelica colonial: Apuntes históricos de la ciudad minera más importante del Virreynato Peruano*, Mariano Patiño Paúl Ortíz (Lima: Huancavelica 21, 2001), 347; Balthasar Ramírez, "Descripción del reyno del Pirú del sitio, temple, provincias, obispados y ciudades; de los naturales, de sus lenguas y traje," in *Juicio de límites entre el Perú y Bolivia*. Víctor Maurtua, Ed. Vol. I (Barcelona: Imprenta de Heinrich y Compañía, 1906), 321; Modesto Bargallo, *La minería y metalurgía en la América española durante la época colonial* (Mexico City: Fondo de Cultura Económica, 195), 39–40; Leon Crosnier, Geologie du Perou. Notice geologique sur les Departments de Huancavelica et d'Ayachcho," in *Annales de mines*. 5th series, Vol. 11 (1852), 36; Joseph Singewald, "The Huancavelica Mecury Deposits, Peru," in *Engineering and Mining Journal*. Vol, 110, No. 11 (1920), 518; Lohmann Villena, *Las minas de Huancavelica*, 11, 13; Mariano Patiño Paúl Ortíz, *Huancavelica colonial: Apuntes históricos de la ciudad minera más importante del Virreynato Peruano* (Lima: Huancavelica 21, 2001), 20–21; José A. Hernández, "Antiguedad y actualidad de la Villa Rica de Oropesa," in *Peruanidad*, Vol. III, No. 12 (Lima, 1943), 942; Colin Cooke, Holger Hintelmann, Jay Ague, Richard Burger, Harald Biester, Julian Sachs and Daniel Engstgrom, "Use and legacy of mercury in the Andes," in *Environmental Science and Technology*, Vol. IV7, No. 9 (May, 2013), 4181–4184; Colin Cooke, Prentiss Balcom, Harald Biester and Alexander Wolfe, "Over three millennia of mercury pollution in the Peruvian Andes," in *Proceedings of the National Academy of Sciences*. Vol. I02, No. 22 (June, 2009), 8830; Arturo Ruiz Estrada, *Arqueología de la ciudad de Huancavelica* (Lima: Servicios deArtes Gráficas, 1977), 30, 37; Georg Petersen, *Mining and Metallurgy in Ancient Perú*. William E. Brooks, Trans. Special Paper 467 (Boulder, CO: The Geological Society of America, 2010), 29; Brooks, *Peru Mercury Inventory*, 3.

21   Canto de Andrade, 304; Lohmann Villena, *Las minas de Huancavelica*, 29–30, 49–50, 63.

of argentiferous ores, some scholars posit that mercury was refined in the region before the arrival of the Spanish.[22] For example, Brooks argues that the similar quantities of residual mercury in pre-Columbian gold artefacts and modern gold produced by mercury amalgamation (on average about ten parts per million) reflect the use of this technique in pre-Hispanic times. Moreover, the sheer amounts of pre-Columbian gold beg the question as to how it was produced in the absence of mercury amalgamation. It is possible, however, that the residual contamination of such artefacts resulted from contact with mercury from natural sources or derived from placer gold which was then melded by fire. This would burn off most of the associated mercury, leaving trace amounts similar to that encountered in amalgamated gold.[23]

The Historiography of Huancavelica

Recent decades have seen an increasing number of studies on the history of the city and region of Huancavelica, although most are in Spanish. Reflecting the importance of mercury to Spain's colonial enterprise, many works concern Huancavelica's colonial period and emphasize administrative, labor

---

22   Cobo, 150; Ruiz Estrada, 50; Government of Peru, *Almanaque de Huancavelica*, (Lima: Instituto Nacional de Estadística e Informática, 2002), 17; Petersen 29; Donald Noble and Cesar Vidal. "Association of Silver with Mercury, Arsenic, Antimony, and Carbonaceous Material at the Huancavelica District, Peru," in *Economic Geology*, Vol 85, No. 7 (1990), 1645. Cooke et al., in "Over three millennia," 8833, indicate that refining began after the region fell under Inca domination, and assert that "at least some emissions must have been in the form of gaseous HGo (or possibly reactive HG2)." This could result from pre-Hispanic cinnabar smelting operations, or the release of gaseous mercury as a result fo smelting other ores, such as silver. They also indicate that given the margin of error of radio carbon dating, which can be 200 years, "We cannot determine if the initial rise in Hg pollution occurred during the latest stages of the Inca Empire or shortly after Hispanic conquest of the Andes." See Cooke et al. "Use and legacy of mercury," 4185. Regarding rumors of pre-colonial retorts in Huancavelica, see Mariano de Rivero y Ustariz, *Memoria Sobre El Rico Mineral De Azogue De Huancavelica*. Vol. II (Lima: J.M. Masías, 1848), 1. See also Rafael Larco Hoyle, *Los Mochicas*. Vol. II (Lima: Museo Arqueológico Rafael Larco Herrera, 2001), and Augusto Cabrera La Rosa, "Situación actual de la minería del mercurio en el Perú," in *Minería y Metalurgia: Boletín official de minas, metalurgia y combistibles* (Madrid: Veritas, 1954), 3–12.

23   William Brooks, Gabriela Schworbel and Luis Enrique Castillo, "Amalgamation and Small-scale Gold Mining in the Ancient Andes", in *Mining and Quarrying in the Ancient Andes: Sociopolitical, Economic and Symbolic Dimensions*. Nicholas Tripcevich and Kevin Vaughan, Editors (New York: Springer Publishing, 2013), 214, 218–219, 223–225.

and health issues, as well as Bourbon efforts to increase quicksilver production through technological improvements. Urban history, the productive potential of the mines, and legacy contamination issues round out the major areas of research. Adrian Pearce offers a detailed review of the literature and available primary sources of colonial Huancavelica, in "Huancavelica, 1563–1824: History and Historiography,"[24] while Rommel Plasencia Soto and Fernando Cáceres Ríos offer an annotated bibliography of diverse works on the city and region in *Bibliografía de Huancavelica*.[25]

There are few studies of the region of Huancavelica prior to the Spanish conquest, however Rubén Darío Espinoza Gonzales offers an overview of the department's archaeological sites in "Una vision de la Arqueología de Huancavelica."[26] With a narrower focus on the city's prehistory is the work of Arturo Ruiz Estrada, titled *Arqueología De La Ciudad De Huancavelica*.[27] Foundational works on the town and Santa Bárbara mine include Arthur Whitaker's pioneering *The Huancavelica Mercury Mine*,[28] which covers the eighteenth century. This was followed in 1948 by Guillermo Lohmann Villena's *Las minas de Huancavelica en los siglos XVI y XVII*,[29] which spans the sixteenth and seventeenth centuries. Both works are, however, largely institutional and administrative histories, detailing the various governors and mercury production contracts in the town. While these two works offer solid and detailed accounts of royal policies, civil administration and quicksilver production, Lohmann Villena's contribution is attenuated by the racist views expressed in the work.[30]

The highly toxic nature of mercury production in Huancavelica has received increasing attention in recent years. Octavio Puche discusses early seventeenth century environmental health issues in "Influencia de la legislación minera, del laboreo, así como del desarrollo técnico y ecnonómico, en el estado y producción de las minas de Huancavelica, durante sus primeros tiempos,"[31] while José Sala Catala focuses on conditions in the mines in the first half of the eighteenth century in "Vida y muerte en la mina de Huancavelica durante la primera mitad

24   *Colonial Latin American Review*, Vol. 22, No. 3 (December, 2013): 422–440.

25   Lima: Universidad Nacional Mayor de San Marcos, 1996.

26   In Arqueología y Desarrollo: Experiencias y posibilidades en el Perú. Luis Valle Alvarez, Editor (Trujillo, Peru: Ediciones sian, 2010), 67–78.

27   Lima: Servicios de Artes Gráficas, 1977.

28   Boston: Harvard University Press, 1941.

29   Lima: Pontífica Universidad Católica del Perú, 1999.

30   Lohmann Villena, *Las minas de Huancavelica*, 3, 10, 11, 104, 242, 416, 437, 305.

31   *Minería y metalurgia: Intercambio tecnológico y cultural entre América y Europa durante el período colonial español*. Manuel Castillo Martos, Ed. (Seville: Muñoz Moya y Montraveta, Editores, 1994), 437–482.

del siglo XVIII."[32] Kendall Brown has also deepened our understanding of the public health aspects of mercury production in Huancavelica, as well as the efforts of Bourbon monarchs to increase production in the mine. His landmark "Workers' Health and Colonial Mercury Mining at Huancavelica, Peru,"[33] offers rich detail and a broad chronological span which charts the evolving conditions in the mine.

Other works have placed Huancavelica within the broader context of Andean silver production. These include Gwendolyn Cobb's *Potosí y Huancavelica, bases económicas, 1545–1640*,[34] which examines the vital role played by Huancavelica's mercury in Potosí's silver production, and her "Supply and Transportation for the Potosi Mines, 1545–1640."[35] More recently, in *Mercury, Mining and Empire: The Human and Ecological Cost of Colonial silver Mining in the Andes*,[36] Nicholas A. Robins examines the environmental effects of, and economic linkages between, colonial mercury refining in Huancavelica and silver production in Potosí. The work utilizes air dispersion modeling to estimate historic mercury vapor concentrations in the air at different levels of production thus illuminating to the health and environmental effects of mineral production.

The historiography of Huancavelica has also benefitted from broader treatments of the city's past. For example, Carlos Contreras offers a colonial history of Huancavelica that takes us to the dawn of the Bourbon era in *La ciudad del mercurio: Huancavelica, 1570–1700*.[37] Mariano Patiño Paúl Ortíz has also published a colonial history of the town, titled *Huancavelica colonial: Apuntes históricos de la ciudad minera más importante del Virreynato Peruano*,[38] and Tulio Carrasco offers an urban chronology in *Cronología de Huancavelica (Hechos, poblaciones y personas)*.[39] Federico Salas Guevara has also authored a synthetic account of the region's history in his two volume, encyclopedic, *Historia de Huancavelica*[40]

The Bourbon era in Huancavelica has been a topic of increasing attention, such as Kendall Brown's penetrating study of royal finances during the War of Spanish Succession (1701–14) and their effect on silver and mercury production, titled "La crisis financiera peruana al comienzo del siglo XVIII, la minería

32   *Asclepio*, Vol. 39 (1987), 193–204.
33   *The Americas*, Vol. 57. No. 4 (April, 2001): 467–496.
34   La Paz: Banco Minero de Bolivia, 1977.
35   *Hispanic American Historical Review*, Vol. 29, No. 1. (1949): 25–45.
36   Bloomington: Indiana University Press, 2011.
37   Lima: Instituto de Estudios Peruanos, 1982.
38   Lima: Huancavelica 21, 2001.
39   Lima: Companía de Minas Buenaventura, 2003.
40   Lima: Compañía de Minas Buenaventura, 2008.

de plata y la mina de azogue de Huancavelica."[41] Adrian Pearce also expands the historiographical horizon of Bourbon reform efforts in his insightful "Huancavelica 1700–1759: Administrative Reform of the Mercury Industry in Early Bourbon Peru."[42] Other, more narrowly focused, administrative studies enhance our understanding of the Bourbon era in Huancavelica. These include Serena Fernández Alonso's examination of the administration of Governor José de Santiago-Concha y Salvatierra, Marqués de Casa Concha, and efforts to boost production in the early eighteenth century, in "Los mecenas de la plata: el respaldo de los virreyes a la actividad minera colonial en las primeras decadas del siglo XVIII. El gobierno del Marqués de Casa Concha en Huancavelica, 1723–1726."[43] The work of Miguel Molina Martínez has illuminated Antonio de Ulloa's term as governor there in both *Antonio de Ulloa en Huancavelica*,[44] and "Tecnica y laboreo en Huancavelica a mediados del siglo XVIII."[45] Other works focusing on the Bourbon era include Alejandro Reyes Flores' "Huancavelica, 'Alhaja de la Corona': 1740–1790,"[46] and Mervyn Lang's study of the devastating 1786 collapse of the mine in "El derrumbe de Huancavelica en 1786. Fracaso de una reforma bourbónica."[47]

Several studies have chronicled the largely successful Bourbon efforts to improve technology and production in the Huancavelica's mines. Most notable in this regard are the publications of Kendall Brown, such as "La recepción de la tecnología minera española en las minas de Huancavelica, siglo XVIII,"[48] "Los cambios tecnológicos en las minas de Huancavelica, siglo XVIII,"[49] and "El ingeniero Pedro Subiela y el desarrollo tecnológico en las minas de Huancavelica (1786–1821)."[50] Vicente Palacio Atard also probes later Bourbon efforts to boost production in "El asiento de la mina de Huancavelica en 1779."[51]

---

41   *Revista de Indias*, Vol. XLVIII, Nos. 182–183 (1988): 349–381.

42   *Hispanic American Historical* Review, Vol. 79, No. 4 (Nov. 1999), 669–702.

43   *Revista De Indias*. Vol. 60, No. 219 (Madrid: 2000), 345–371.

44   Granada, Spain: University of Granada, 1995.

45   *Europa e Iberoamérica: Cinco siglos de Intercambios*, Vol. II. María Justina Sarabia Viejo, et al., Eds. (Seville, Spain: Asociación de Historiadores Latinoamericanistas Europeos/ Consejeria de Cultura y Medio Ambiente, 1992), 395–405.

46   *Ensayos en ciencias sociales*. Julio Mejía Navarrete, Ed. (Lima: Universidad Nacional Mayor de San Marcos, 2004).

47   *Histórica*, Vol. 10, No. 2 (Dec., 1986): 213–226.

48   *Saberes andinos: ciencia y tecnología en Bolivia, Ecuador y Perú*, Marcos Cueto, Ed. (Lima: Instituto de Estudios Peruanos., 1995).

49   *Hombres, técnica, plata: minería y sociedad en Europa y América, siglos XVI–XIX*, Julio Sánchez Gómez and Guillermo Mira Delli-Zott, Eds. (Seville: Aconcagua Libros, 2000).

50   *Histórica* Vol. 30, No. 1 (July, 2006), 165–184.

51   *Revista de Indias* Vol. 5 (1944), 611–630.

Late colonial efforts to improve the supply of mercury in the Andes is
• the topic of María Dolores Fuentes Bajo's "El azogue en las postrimerías del
Perú colonial."[52] With the upheaval of the wars for independence came new
challenges concerning the provision of mercury in both the Andes and New
Spain, a topic explored by Kendall Brown in "La distribución del mercurio a
finales del periodo colonial, y los trastornos provocados por la independencia
hispanoamericana."[53]

Among the more controversial topics concerning Huancavelica is the nature and extent of the *mita*, or Indian draft labor system, which was imposed
there. Silvio Zavala provides a detailed, institutional, account of the mita in *El
servicio personal de los indios en el Perú*,[54] while Jorge Basadre focuses specifi-
cally on the Huancavelica mita, and the debate surrounding it, in "El Régimen
de la Mita."[55] Paulino Castañeda Delgado explores the moral dilemmas posed
by utilizing draft indigenous labor in both "El tema de las minas en la ética
colonial española,"[56] and "Un capítulo de ética Indiana española: los trabajos
forzados en las minas."[57] Other treatments of the mita include Donald Wiedner's "Forced Labor in Colonial Peru,"[58] and the work of Luis Basto Girón, in
"Las mitas de Huamanga y Huancavelica."[59]

Isabel Povea Moreno offers a more recent study of labor issues in "Entre
la retórica y la disuación. Defensores e impugnadores del sistema mitayo en
Huancavelica y en las Cortés de Cádiz,"[60] while in *Minería y reformismo borbónico en el Perú. Estado, empresa y trabajadores en Huancavelica 1784–1814*,[61]
she expands the discussion beyond labor to explore administrative, technical
and health issues in the late Bourbon era. Offering a counterpoint to much of
what has been written about the mita, Barbara Bradby impugns the widely
held perception of the rigors of work in the Santa Bárbara mine by emphasiz-
ing the role of wage, as opposed to mita, labor there, in "The 'Black Legend' of

---

52   *Revista de Indias*. Vol. XLVI (January–June, 1986), 75–105.

53   *Minería colonial Latinoamericana*, Dolores Avila, Inés Herrera and Rina Ortíz, Eds. (Mex-
      ico City: Instituto Nacional de Antropología e Historia, 1992), 155–160.

54   Mexico City: El Colegio de Mexico, 1978.

55   *Letras*, Vol. III (Lima: 1937): 325–364.

56   *La minería hispana e iberoamericana*. Vol. I. na. (León, Spain: Catedra de San Isidoro,
      1970), 333–354.

57   *Anuario de Estudios Americanos*, Vol. XXVII, (1970), 815–916.

58   *The Americas*, Vol.16, No. 4, (April, 1960), 357–383.

59   *Perú Indígena* Vol. 5, No. 13 (December, 1954): 215–242.

60   *La Constitución Gaditana de 1812 y sus repercusiones en América*, Alberto Gullón Abao and
      Antonio Gutiérrez Escudero, Eds. (Cadiz: Universidad de Cadiz, 2012), 201–211.

61   Lima: Instituto de Estudios Peruanos, 2014.

Huancavelica: The mita debates and opposition to wage-labor in the colonial mercury mine."[62] Labor is also the point of departure for the study by Adrian Pearce and Paul Jeggarty, who explore how the convergence of workers from regions to the south of Huancavelica affected the dialect of Quechua spoken in the region in "'Mining the data' on the Huancayo-Huancavelica linguistic frontier."[63]

In contrast to the abundance of works that concern the colonial era, there are few studies concerning Huancavelica in the nineteenth century. Most concern informal mining and efforts to resume industrial mercury production in the region. For example, late colonial artisanal mining is the focus of Isabel Povea Moreno's "Los buscones de metal. El sistema de pallaqueo en Huancavelica (1793–1820),"[64] while Carlos Contreras and Ali Díaz expand the chronological scope by delving into the long-neglected, and misunderstood, history of nineteenth century mercury production in Huancavelica in *Los intentos de reflotamiento de la mina de azogue de Huancavelica en el siglo XIX*.[65]

Most nineteenth, and early twentieth, century studies of Huancavelica were, however, prepared by scientists, geologists and mining engineers evaluating the potential of Huancavelica's regional cinnabar deposits and other natural resources, and the most effective means to exploit them. One of the earliest, and most prolific was the indefatigable Italian naturalist and geographer Antonio Raimondi (1826–1890). In his many travels throughout Peru, he not only explored the mines of Santa Bárbara but also described the flora, fauna, geography and geology of the region. Much of this work is contained in his six volume *El Perú*[66] and "Huancavelica y mina de azogue, año 1862."[67] In the mid-eighteenth century Mariano de Rivero y Ustariz provided an oft-cited estimate of Huancavelica's quicksilver production and detailed descriptions of the refining process in *Memoria Sobre El Rico Mineral De Azogue De Huancavelica*.[68] His work was followed by that of C.E. Hawley, who discussed the geology of the region and smelting processes in "Notes on the Quicksilver Mine of Santa

---

62   *Hombres, técnica, plata: minería y sociedad en Europa y América, siglos XVI–XIX*, Julio Sánchez Gómez and Guillermo Mira Delli-Zotti, Eds., (Seville: Aconcagua Libros, 2000), 227–257.

63   *History and Language in the Andes*. Paul Heggarty and Adrian J. Pearce, Eds. (New York: Palgrave MacMillan, 2011).

64   *Anuario de Estudios Americanos*, Vol. 69, No. 1 (2012): 109–38.

65   Lima, NP, 2007.

66   Lima: Imprenta del Estado, 1874.

67   *Notas de viajes para su obra "El Peru."* Vol. 3 (Lima: Imprenta Torres Aguirre, 1945), 276–289.

68   Lima: J.M. Masías, 1848.

Barbara, Peru,"[69] while less than two decades later, Eduardo de Habich published "Industria de azogue,"[70] in which he situates the mine's potential in a changing global market for mercury. The same journal also published, in 1889, Juan Torrico y Mesa's description of the Santa Bárbara mine and region in "Excursión al departamento de Huancavelica"[71] and "El azogue de Huancavelica"[72] in which Santiago Loveday examines the ore quality and potential of the cinnabar deposits. The year before, William Hadley published his "Report on the quicksilver mines of Huancavelica,"[73] as did Augusto Tamayo his "Mina de cinabrio 'Santa Bárbara' en Huancavelica,"[74] both of which were oriented to promoting renewed production there. Similarly, Augusto F. Umlauff examined Huancavelica's geology and production in "El Cinabrio De Huancavelica,"[75] as did Carlos Jiménez in "Estadistica Minera en 1917."[76]

Building on such works, in the early twentieth century Pedro Arana prepared a concise history of mercury production in Huancavelica with an eye on its revitalization in *Las minas de azogue del Perú*,[77] while soon after Enrique Dueñas provided a valuable description of the city and geology of the region in "Fisionomia minera de las provincias de Tayacaja, Angaraes y Huancavelica."[78] On the heels of these authors was Lester Strauss, who in "Quicksilver at Huancavelica, Peru,"[79] studied the ore quality and the potential of renewed mercury production. In 1922, Eduard Berry and Joseph T. Singewald also examined the geology and state of the mine in *The Geology and Paleontology of the Huancavelica Mercury District*.[80] Complementing previous geological studies is James Wise and Jean Féraud's more recent "Historic maps used in new geological and engineering evaluation of the Santa Bárbara mine, Huancavelica mercury district, Peru,"[81] in which they discuss regional geology and examine colonial and more recent maps to better understand historic production issues.

---

69 *American Journal of Science*, 2nd series. Vol. 45 (1868), 5–9.
70 *Boletín de Minas, Industrias y Construcciones* Vol. 2 (Lima, 1885), 11–13.
71 *Boletín de minas, industrias y construcciones*, Vol. 5, No. 12 (Lima: 1889): 889–893.
72 *Boletín de minas, industrias y construcciones*,Vol. 20 (Lima: 1905): 82–84.
73 *Boletín del Ministerio de Fomento*, Vol. 2, No. 1 (Lima, 1904), 43–44.
74 *Boletín del Ministerio de Fomento*, Vol. 2, No. 1 (Lima, 1904), 38–43
75 *Boletin del Cuerpo de Ingenieros de Minas del Peru*, No. 7 (Lima: Librería Escolar é Imprenta de E. Moreno, 1904).
76 *Boletín del cuerpo de ingenieros de minas del Peru*. No. 95. Lima: Imprenta Americana, 1919.
77 Lima: Imprenta de "El Lucero," 1901.
78 *Boletín del cuerpo de ingenieros de minas del Peru*. No. 62. Lima: El Lucero, 1908.
79 *Mining and Science Press*. Vol. 99 (October 23, 1909), 561–566.
80 Baltimore: Johns Hopkins Press, 1922.
81 *De Re Metallica*, Vol. 4 (2005), 15–24.

Although Huancavelica was usually not their final destination, nineteenth century traveler's accounts provide valuable descriptions of the town during this period. These include such works as Heinrich Witt's *Diario y observaciones sobre el Perú (1824–1890)*,[82] José María Blanco's account of President Orbegoso's 1834 visit to the town, in *Diario de viaje del Presidente Orbegoso al sur del Perú.*,[83] and L. Carranza's "De Huanta á Lima por el camino de Huancavelica – Año de 1866."[84]

Sources for twentieth century studies of the city and region include many publications by the government of Peru, such as censuses, almanacs, catalogues of the natural resources in the region, and public health studies. Among the latter, and useful for understanding contemporary public health challenges in Huancavelica, is the *Encuesta demográfica y de salud familiar 2000: Huancavelica.*[85]

Offering a collection of the region's natural resources is the *Inventario Y Evaluación De Los Recursos Naturales De La Zona Altoandina Del Perú: Reconocimiento, Departamento De Huancavelica.*[86] The importance of tubers in the Andean diet has led to several studies of their variety, evolution and the risks they face in genetic diversity. Leading the way in this area has been the work of Stef De Haan and his colleagues, in works such as the *Catálogo De Variedades De Papa Nativa De Huancavelica, Perú,*[87] "Land Use and Potato Genetic Resources in Huancavelica, Central Peru,"[88] and "Multilevel Agrobiodiversity and Conservation of Andean Potatoes in Central Peru: Species, Morphological, Genetic, and Spatial Diversity."[89]

Huancavelica is one of Peru's most underdeveloped departments, and, reflecting this, many late-twentieth century studes concern the region's economic challenges and efforts to overcome them. The completion in 1926 of the railroad between Huancavelica and Huancayo was the source of much, still unfulfilled, hope for economic development and regional integration among Huancavelicanos. Miguel Pinto Huaracha and Alejando Sánchez Salinas detail the political debates and financial challenges encountered during the construction of this railroad in *Las Rutas Del Café Y El Trigo: Los Ferrocarriles De*

82   Kika Garland de Montero, Trans. (Lima: Corporación Financiera de Desarrollo, s.a., 1987).

83   2 Vols. Felix Denegri Luna, Ed. (Lima: PUCP-IRA 1974).

84   *Boletín de la Sociedad Geográfica de Lima.* No. 5 (September, 30, 1895), 176–187.

85   Vol. 8. (Lima: Instituto Nacional de Estadística e Informática, 2001).

86   2 Vols (Lima: Oficina Nacional de Evaluación de Recursos Nacionales, 1984).

87   Lima: Centro Internacional de la Papa, 2006.

88   *Journal of Land Use Science.* Vol. 5, No. 3 (2010): 179–195.

89   *Mountain Research and Development.* Vol. 30, No. 3 (2010): 222–231.

Chanchamayo Y Huancavelica 1886–1932[90] while Amador Mendoza Ruiz offers a somewhat more romantic account in *Crónicas Del Tren Macho.*[91]

Although Huancavelica no longer produces mercury on an industrial scale, mining continues to be a regional economic motor. Studies of the late-twentieth century mining sector and regional development challenges include the work of Daffos Aste, in *Minería Y Desarrollo Regional: Los Casos De Junín Y Huancavelica, 1970–86,*[92] Carmen López Cisneros' *Huancavelica Ya No Es Tierra Del Mercurio,*[93] and Oswaldo Rivas Berroca's *Huancavelica: Bases Para El Desarrollo Económico Y Social Del Departamento Huancavelica.*[94] The origins and objectives of mining labor in the region form the basis of Heraclio Bonilla and Carmen Salazar's *Formación del mercado laboral para el sector minero. (La experiencia de Huancavcelica, Perú, 1950–1978),*[95] as well as Henri Favre's "La industria minera de Huancavelica en la década de 1960."[96]

Legacy contamination, or pollution from previous mining activities, is an area of increasing interest among researchers of Huancavelica. For example, Colin Cooke and colleagues have examined sediment core samples from Peruvian lagoons in "Over three millennia of mercury pollution in the Peruvian Andes."[97] Legacy mercury contamination in the city of Huancavelica, and the risks it poses to residents, has been the topic of works by Nicholas A. Robins, Nicole Hagan and colleagues, such as "Mercury Production and Use in Colonial Andean Silver Production: Emissions and Health Implications,"[98] "Estimations of Historical Atmospheric Mercury Concentrations from Mercury Refining and Present-Day Soil Concentrations of Total Mercury in Huancavelica Peru,"[99] "Residential Mercury Contamination in Adobe Brick Homes in Huancavelica, Peru,";[100] and "Speciation and bioaccessibility of mercury in adobe bricks and dirt floors in Huancavelica, Peru."[101]

90   Lima: Seminario de Historia Rural Andina, Universidad Nacional Mayor de San Marcos, 2009.

91   Lima, Perú: Niger Editions, 1998.

92   Lima, Perú: Fundación Friedrich Ebert, 1989.

93   Lima: Fundación Friedrich Ebert, 1983.

94   Lima: Editorial Monterrico, 1989.

95   Lima: PUCP, 1983.

96   *Boletín de Lima,* No. 161 (2010), 85–89.

97   *Proceedings of the National Academy of Sciences.* Vol. 102, No. 22 (June, 2009), 8830–8834.

98   *Environmental Health Perspectives,* Vol. 120, No. 5 (May, 2012), 627–631.

99   *Science of the Total Environment.* Vol. 426, No. 1 (June, 2012), 146–154.

100   *PLoS ONE,* Vol. 88, No. 9 (September, 2013), 1–9.

101   *Environmental Geochemistry and Health* Vol. 37, No. 4 (August, 2014).

Today, the city of Huancavelica has about 49,000 residents, many of whom live povertous lives in earthen homes contaminated with mercury, arsenic and lead.[102] Its remoteness, small size and poverty belie the magnitude of its global economic contributions. The city's history is inseparable from that of modern capitalism, as the quicksilver which flowed from it was the vital ingredient which allowed Andeans to produce quantities of silver which enabled the development of the world's first sustained global trade networks, spurred the Industrial Revolution, and yielded today's global economy. That story, however, has origins as old as humanity, and, as we shall see, the cinnabar deposits of Huancavelica were revered long before the arrival of the Spanish.

---

102   Government of Peru, *Censos Nacionales 2007, XI De Población Y VI De Vivienda: Resultados Definitivos*, 941; Government of Peru, *Estado de la Población Peruana, 2014*, 7; Government of Peru, Instituto Nacional de Estadística, Sistema de Consulta de Principales Indicadores de Pobreza, Mapa de Pobreza, http://censos.inei.gob.pe/Censos2007/Pobreza/; Thoms, Robins, Ecos, Brooks and Espinoza Gonzales, *Results of June–July, 2015 Field Study*, 6, 11, 12; Thoms, Robins, Ecos and Brooks. *Results of June/July 2016 Assessment of Soil and Fish*, 9-10. See also Tables 2, 3 and 8 in appendix.

CHAPTER 1

Huaca-villca: From Geological Formation to Spanish Conquest

The city of Huancavelica emerged as a result of volcanic activity, in more ways than one.[1] The Andes were formed primarily by uplift as a result of the Nazca plate moving below, or subducting under, the South American plate beginning in the Triassic period, 200 to 250 million years ago. Such subduction also led to volcanism, which further added to mountain growth in the Andes, as did hydrothermal mineralization. In the department, or state, of Huancavelica, the result of these processes is that the Andes consist primarily of limestone and sandstone.[2]

Both can be impregnated with cinnabar, which acts as a filler of sorts, infiltrating the gaps between the sand particles in sandstone, and in the fissures of sandstone, limestone and igneous rock. Cinnabar is, however, usually found in greater abundance in sandstone, which can also be laced with native mercury, sulphur, and other minerals, such as lead in the case of Huancavelica. Arsenic is also associated with cinnabar deposits, and in some cases its toxic vapors would limit or complicate excavation in Huancavelica's mines.[3] Although

---

1   For a discussion of the geology of the region, see Berry and Singewald, 28–42; Augusto F. Umlauff, "El Cinabrio De Huancavelica," in *Boletin del Cuerpo de Ingenieros de Minas del Peru*. No. 7 (Lima: Librería Escolar é Imprenta de E. Moreno, 1904), 33–41; Enrique Dueñas, "Fisionomia minera de las provincias de Tayacaja, Angaraes y Huancavelica," in *Boletín del cuerpo de ingenieros de minas del Peru*. No. 62 (Lima: El Lucero, 1908), 150–162. For other works on the topic see Rommel Plasencia Soto and Fernando Cáceres Ríos, *Bibliografía de Huancavelica* (Lima: Universidad Nacional Mayor de San Marcos, 1996), 31–46.
2   Berry and Singewald, 51; James Wise and Jean Féraud, "Historic maps used in new geological and engineering evaluation of the Santa Bárbara mine, Huancavelica mercury district, Peru," in *De Re Metallica*, Vol. IV (2005), 16; W.C.F. Purser, *Metal Mining in Peru, Past and Present* (New York: Praeger Publishers, 1971), 19; Brading and Cross, 547; E.H. McKee, E.H., D.C. Noble, and Cesar Vidal, "Timing of Volcanic and Hydrothermal Activity, Huancavelica Mercury District, Peru," in *Economic Geology and the Bulletin of the Society of Economic Geologists*. Vol. VIII1, No. 2 (1986), 490.
3   Emetherio Ramírez de Arellano. Huancavelica, April 3, 1649. AGI, Lima, 279–284, 1; Robert Yates, Dean F. Kent, and Concha J. Fernández, *Geology of the Huancavelica Quicksilver District, Peru* (Washington, D.C.: U.S. G.P.O, 1951), 1; Wise and Féraud, 16; C.E. Hawley, "Notes on the Quicksilver Mine of Santa Barbara, Peru," in *American Journal of Science*, 2nd series, Vol 45 (1868), 5–7; Lester Strauss, "Quicksilver at Huancavelica, Peru," in *Mining and Science Press*. Vol. IX9 (October 23, 1909), 562–563; Berry and Singewald, 29–30; Joseph T. Singewald,

cinnabar deposits are found throughout Peru between six and fifteen degrees latitude, the ore contained in the Santa Bárbara and neighboring Chacllata-cana mounts near of the city of Huancavelica are the most significant, and, prior to exploitation, were among the largest in the world.[4]

The richness, or grade, of cinnabar ore is often inversely proportional to the depth at which it is located. In the case of the Santa Bárbara, the largest deposit is concentrated near its summit in an area around 550 meters long and about ninety wide. This, in turn, is only a part of the much larger mineralization belt upon which the city rests. Running in a northwest to southeast direction, the belt is about sixty kilometers long and three wide and harbors over forty cin-nabar deposits.[5] Most, however, appear to be concentrated between Ventanilla to the south of the city and Yana-Padre to the north. This belt dates back six-teen million years, although most of the cinnabar deposits were formed about seven million years ago. Beyond the cinnabar mineralization belt, the region of

"The Huancavelica Mecury Deposits, Peru," in *Engineering and Mining Journal.* Vol, 110, No. 11 (1920), 518, 520–521; D'Itri, 118; Government of Peru, *Inventario Y Evaluación De Los Recursos Naturales De La Zona Altoandina Del Perú: Reconocimiento, Departamento De Huancavelica.* Vol. I (Lima: Oficina Nacional de Evaluación de Recursos Nacionales, 1984), 81; McKee et al, 489–490; José Parés y Franqués, *Catástrophe morboso de las minas mercuriales de la villa de Almadén del azogue (1778).* Alfredo Menéndez Navarro, Ed. (Cuenca, Spain: Ediciones de la Universidad de Castilla – La Mancha, 1998), 237; Antonio G. Gastelumendi, *Huancaveli-ca como región productora de mercurio.*(Lima: Imp. Torres Aguirre, 1920), 41; Lohmann Vil-lena, *Las minas de Huancavelica,* 183; Craig Scribner, *To Air Is Human: Oppressive Conditions in Huancavelica's Mercury Mine from 1600–1616 and Efforts to Remedy Them.* Honors Thesis (Brigham Young University, 1995), 10; Dueñas, 161; Hawley, 7; Contreras and Díaz, 16; Kendall Brown, "Workers' Health and Colonial Mercury Mining at Huancavelica, Peru," in *The Ameri-cas,* Vol. V7. No. 4 (April, 2001), 472; Noble David Cook, *Demographic Collapse: Indian Peru, 1520–1620* (Cambridge: Cambridge University Press, 1981), 205. Often encountered in conjunc-tion with cinnabar is bitumen, a hydrocarbon, galena, or lead sulfide, arsenopyrites, or iron arsenic sulfide, arsenic sulfides such as realgar and orpiment, silica, sulphur and iron pyrites.

4   Pedro Fernández de Castro y Andrade, Conde de Lemos, "Refiere lo obrado en poco más de un áno de servicio." January 20, 1669. In *Los virreyes españoles en America durante el gobierno de la casa de Austria. Perú.* Vol. IV. Lewis Hanke and Celso Rodríguez, Eds. (Madrid: IMNASA, 1978), 275; Berry and Singewald, 9–10; Wise and Féraud, 16; D'Itri, 118; McKee, 489.

5   Hawley, 6; Berry and Singewald, 42; Wise and Féraud, 16; Pedro Arana, *Las minas de azogue del Perú* (Lima: Imprenta de "El Lucero," 1901), 2–3; Strauss, 562–63; Yates et al., 2; Dueñas, 160; Singewald, 518; Santiago Loveday, "El azogue de Huancavelica," in *Boletín de minas, industrias y construcciones.* Vol. IIo (Lima: 1905), 83; Gastelumendi, 40; Juan Torrico y Mesa, "Excursión al departamento de Huancavelica," in *Boletín de minas, industrias y construcciones,* Vol. V, No. 12 (Lima: 1889), 891; Augusto Tamayo, "Mina de cinabrio 'Santa Bárbara' en Huancavelica," in *Boletín del Ministerio de Fomento,* Vol. II, No. 1 (Lima, 1904), 39; Rivero y Ustariz, 8–9; Arana, 2, 11, 44; Hawley, 7; Berry and Singewald, 19–20; Patiño Paúl Ortíz, 100–101.

Huancavelica is endowed with numerous other minerals. These include gold, silver, antimony, pyrites, zinc, wolfram, lead, as well as coal, phosphates, plaster, limestone, marble and arsenic sulfides such as orpiment and realgar.[6]

Unlike argentiferous ore where deposits usually occur as veins, cinnabar is commonly found in pockets. As a result, its extraction has a greater tendency to undermine the geological stability of the surrounding area, especially in porous rocks such as limestone and sandstone. Such risks can be minimized if natural or other supports are left or created. In the case of Huancavelica, such *puentes*, or rock archways, and *estribos*, or columns, were very rich in cinnabar. Despite repeated prohibitions to the contrary, colonial miners routinely excavated them, risking lives in a subterranean game of Russian roulette that often ended abruptly in a lethal cave-in.[7]

Geological Gifts and Curses

Smelting is the prevalent method of producing mercury, although today it is also produced as a byproduct in the refining of silver. At Santa Bárbara, mercury is also encountered in a naturally occurring elemental form in sedimentary deposits a meter or so below the surface. Because the city of Huancavelica straddles both the Ichu River and the cinnabar mineralization belt, native mercury is often encountered there. While it is possible that some subsurface quicksilver found in the city is the legacy of refining, the three foot or so depth at which it is usually encountered suggests that it is the result of

---

6   Rivero y Ustariz, 8–9; Arana, 2, 11, 44; Rivero y Ustariz, 8–9; Hawley, 7; Berry and Singewald, 19–20; Patiño Paúl Ortíz, 100–101; Parés y Franqués, 237; Gastelumendi, 41; Lohmann Villena, *Las minas de Huancavelica*, 183; Scribner, 10; Dueñas, 161; Contreras and Díaz, 16; Brown, "Worker's Health," 472; Cook, *Demographic Collapse*, 205; McKee, 490; Government of Peru, *Censo Nacional De Población Y Ocupación, 1940.* Vol. VI. (Lima: Dirección Nacional de Estadística y Censos,1949), viii; Iden, *Centros Poblados: Sexto Censo Nacional De* Población, Vol. II., 260; Iden, *Inventario Y Evaluación De Los Recursos Naturales*, Vol. I, 81; Carlos P. Jiménez, "Estadistica Minera en 1917," in *Boletín del cuerpo de ingenieros de minas del Perú.* No. 95 (Lima: Imprenta Americana, 1919), 21; Federico Salas Guevara, *Historia De Huancavelica* Vol. I (Lima: Compañía de Minas Buenaventura, 2008), 1, 4. See also Tulio Carrasco, *Cronología de Huancavelica (Hechos, poblaciones y personas)* (Lima: Companía de Minas Buenaventura, 2003), 384–434.

7   "Memorial del capitán Don Pedro Gutíerrez Calderón de algunas advertencias considerables al servicio de Dios Nuestro Señor y de su Real Magestad," Lima, May 2, 1623, AGI, Lima, 154,1; Hawley, 6; Lohmann Villena, *Las minas de Huancavelica*, 180; Pearce, "Huancavelica 1700–1759, 677–678"; Scribner, 5; Pearce, "Huancavelica 1700–1759," 678.

cinnabar reacting with acidic water or other natural processes which yield native mercury. Cause, Huancavelica's residents have long encountered such deposits. For example, in addition to the account with which this book opens, in the early eighteenth century a residential excavation in the town yielded several tons of native mercury, prompting many residents to eagerly prospect for more in the area.[8] In the early twentieth century, excavations for a building foundation on the city's central plaza yielded almost one and one-half metric tons of liquid mercury.[9] Even today, residents often report encountering native mercury at depths of about a meter when digging building foundations, gathering soil to construct earthen homes, installing drainage or improving roads.[10]

The geological processes which produced Huancavelica's cinnabar deposits have also endowed the city with mineral-infused thermal springs. The concentrations of calcium carbonate and iron oxide in these waters are so high that, when poured into forms and evaporated, it yields travertine, a form of limestone. Natural travertine deposits reaching several meters of thickness, resulting from centuries of accretion from thermal springs, are also encountered along the Ichu River downstream from the city. Travertine's potential was recognized by the early colonial inhabitants, who found that it offered an easily worked and durable construction material. Many government buildings, churches and elite homes were constructed of this stone, and in some cases remain to this day. Beyond being a source of travertine for the city's residents, the local mineral springs have served for hundreds of years as public baths, and continue to do so in the San Cristóbal neighborhood.[11]

Although Huancavelica's geological characteristics led to the city's creation, they can also be a destructive force. Because the city is situated on the South American tectonic plate, it is vulnerable to earthquakes as the neighboring Nazca plate subducts, or moves beneath it. The town and region have suffered

---

8   Goldwater, 49; Berry and Singewald, 43; Raquel Delgado de Castro, *El Despertar De Huan-cavelica* (Lima: C.A. Castrillón, 1927), 73; Singewald, 522; Strauss, 562; Crosnier, 51–52; Berry and Singewald, 43; Delgado de Castro, 73.

9   Gastelumendi, 41.

10   Nicholas Robins, Field Research Notes, August, 2010, July, 2012, June–July, 2015.

11   Berry and Singewald, 51; *D'Itri*, 118; Antonio Raimondi, *El Perú*, Vol.4 (Lima: Imprenta del Estado, 1874), 310; Cantos de Andrade, Murua, 550–551; 305; Raimondi, *El Perú*, Vol. IV, 310–311; Ibid., Vol. V, 39; Tadeo Haenke, *Descripción del Perú* (Lima: Imprenta El Lucero, 1901), 261; Torrico y Mesa, 891; Lohmann Villena, *Las minas de Huancavelica*, 68; 224; José María Blanco, *Diario de viaje del Presidente Orbegoso al sur del Perú*. Vol. I, Felix Denegri Luna, Ed. (Lima: PUCP-IRA 1974), 27; Salas Guevara, *Historia de Huancavelica*, Vol. I, 3, 5–6; Dueñas, 155; Strauss, 562.

many temblors, such as those of 1650, 1687 and 1690 which damaged buildings in the town. In the 1687 event, the towns of Huanta and Angaraes paid an especially high price of damage and death. Later, in 1719, another earthquake rattled buildings and nerves in the town, but caused little damage.[12] Much more destructive was the infamous quake which struck in the late morning of October 28, 1746. Centered about sixty-five kilometers northwest of Lima and with a magnitude of at least 8.4 on the Richter scale, it remains the most severe earthquake in Peru's recorded history. In Huancavelica, it destroyed many buildings, including the San Augustín church. The Dominican convent also sustained damage, while the towers of the Santo Domingo, San Sebastián, Santa Ana and San Juan de Dios churches were toppled. Despite the damage, Huancavelica escaped relatively unscathed. Lima and its neighboring port of El Callao were devastated, with over 1,100 people perishing in the quake itself, and many more from ensuing diseases. On the heels of the tremor, what was left of El Callao was swept away by a tsunami, killing at least 5,000 more people.[13]

A less lethal, but much more frequent, menace to the city and region of Huancavelica is the volatile climate. Sudden deluges, hail, snowstorms and freezes have long tormented the region. While hail can damage crops and often precariously built homes, severe rainstorms often cause landslides and flash flooding in the city's sharp valley. The perils are compounded as such events bathe the city in toxic runoff from the region's mercury, arsenic and lead contaminated soils.[14]

---

12   Melchor Navarra y Rocaful, Duque de la Palata, "Relacion del estado del Perú," in *Memorias de los virreyes que han gobernado el Perú durante el tiempo del coloniaje español*. Vol. II. M.A. Fuentes, Ed. (Lima: Librería Central de Felipe Bailly, 1859), 178; Carrasco, *Cronología de Huancavelica*, 154, 206; Lohmann Villena, *Las minas de Huancavelica*, 449, 457; José Armendaris, Marques de Castel-Fuerte, "Relacion del estado de los reynos del Perú que hace el Excmo. Señor Don José Armendaris, Marqués de Castel-Fuerte, á su successor el Marqués de Villagarcía, en el año de 1736," in *Memorias de los virreyes que han gobernado el Perú durante el tiempo del coloniaje español*. M.A. Fuentes, Ed. Vol. III. (Lima: Librería Central de Felipe Bailly, 1859), 172, Carrasco, 220–221; Serena Fernández y Alonso, "Los mecenas de la plata: el respaldo de los virreyes a la actividad minera colonial en las primeras decadas del siglo XVIII. El gobierno del Marqués de Casa Concha en Huancavelica, 1723–1726," in *Revista De Indias*. Vol. VIo, No. 219 (Madrid: 2000), 353.

13   Salas Guevara, *Historia de Huancavelica*, 210; Charles Walker, *Shaky Colonialism: The 1746 Earthquake-Tsunami in Lima, Peru and its Long Aftermath* (Durham, Duke University Press, 2008), 1–2, 7–8, 11, 69.

14   Tadeo Haenke, *Descripción del Perú* (Lima: Imprenta El Lucero, 1901), 261; Torrico y Mesa, 891; Salas Guevara, *Historia de Huancavelica*, 191; Bryn Thoms and Nicholas Robins, "Remedial Investigation. Huancavelica Mercury Remediation Project. Huancavelica, Peru." Unpublished Manuscript. July, 2015, 7; Thoms, Robins, Ecos, Brooks and Espinoza

Andean Abundance

It is in this environmental context that humans first settled in the region. On the site of what is now the city of Huancavelica, at 3,378 meters feet above sea level, it appears that the first sustained human settlement occurred between around 1,000 BC and 0 AD, although nomadic hunter-gathers probably inhabited the region up to 10,000 years earlier. As sedentary groups emerged, they utilized ceramics for food storage and presentation, and worked bone and metal into tools and weapons. Over time, the vertical archipelago system emerged in which *allyus*, or extended kin groups, would disperse and settle in different microclimates and exchange the goods they produced, enabling all to enjoy a diversity of foods and products.[15]

While the region's residents have long suffered the travails associated with an unstable geology and mercurial weather, they have benefitted from the rich diversity of the region's flora and fauna. The department hosts no less than twenty-five life zones, as determined according to the Holdridge classification system by the average temperature, yearly precipitation, and percent of total plant transpiration and evaporation relative to rainfall.[16] The department of Huancavelica is also geographically diverse, rising from about 1,000 meters above sea level to mountain peaks approaching 5,300 meters, from which snowmelt feeds rivers which drain to both the Pacific and Atlantic oceans. Agave, in addition to cedar, molle (*schinus molle*) and carob trees prosper at lower elevations, while willows and alders grow at higher elevations. These trees, as well as non-native pine and eucalyptus, commonly serve as construction or fencing materials. Nine varieties of corn, as well as quinoa, barley, wheat, oats, fava and over sixty medicinal plants also thrive in the region.[17]

      Gonzales, *Results of June–July, 2015 Field Study,* 6, 11–12; Thoms, Robins, Ecos and Brooks. *Results of June/July 2016 Assessment of Soil and Fish,* 9–10. See also Tables 1, 3, 4 in appendix.

15   Ruiz Estrada, 14, 18, 25, 33; Carrasco, 39; Oswaldo Rivas Berrocal, *Huancavelica: Bases Para El Desarrollo Económico Y Social Del Departamento Huancavelica* (Lima: Editorial Monterrico, 1989), 27; Government of Peru, *Almanaque De Huancavelica,* 15; Iden, *Censo Nacional De Población Y Ocupación, 1940,* Vol. VI, v; Cook, 25; Heraclio Bonilla, "1492 y la población indígena de los Andes," in *Los conquistados: 1492 y la poblacion indígena de las Américas.* Heraclio Bonilla, Robin Blackburn, et al., Eds. (Bogotá: Tercer Mundo Editores/ Facultad latinoamericana de ciencias sociales, 1992,) 108, 110.

16   Mariela Espinoza Flores, *Huancavelica: Rincón De Misterios Y Encantos/ A Spot Full of Mysteries and Charm* (Lima: Compañía de Minas Buenaventura, 2009), 93.

17   Espinoza Flores, 94, 99, 105; Government of Peru. *Censo Nacional De Población Y Ocupación, 1940.* Vol. VI. (Lima, 1949), vii, 3; Iden, *Inventario y evaluación de los recursos naturales de la zona altoandina del Perú: Reconocimiento, Departamento de Huancavelica, Vol. I,*

The largely treeless altiplano derives its characteristic beige color from the prevalence of *ichu* (*stipa ichu*), the spiny straw which fueled the town's mercury furnaces until the twentieth century. Also found on the altiplano is *yareta*, (*azorella yarita*), a very slow growing, dense, highly resinous green plant resembling moss. This too served as a combustible for refining, before overuse led to its near disappearance from the region and ichu was substituted in its place. The short, robust queñua tree (*polylepsis rugulosa*) is also found on the altiplano, yet was also, and in some places still is, overharvested in the region for fuel and construction materials.[18]

Other flora at these higher altitudes include *tola* (*Epidophyllum quadrangulare*), a shrub which will burn while green, *champa estrella* (*Distichia muscoides*), a perennial with a similar appearance to yareta, and *juncus*, a form of rush, which prospers in moist environments. Perhaps one of the most unusual plants in the department is the *Puya Raimondi*, a magnificent bromeliad named after the Italian naturalist Antonio Raimondi. While the plant itself can reach almost three meters, the flower can rise seven meters beyond that. Encountering a bloom is a rare event, as the plant flowers only once in about a century before dying.[19]

One of the most important crops in the region is tubers. Of the eight species of cultivated potato, seven are found in the department of Huancavelica, where they are joined by seven species of wild potato. Of the cultivated species, the department hosts no less than 570 varieties, 144 of which are indigenous. Tubers commonly grown in the region include numerous cultivars of *oca* (*oxalis tuberosa*), *mashua* (*tropaeolum tuberosum*), sweet potato (*Ipoea batatas*), and *olluco* (*Ullucus tuberosus*), the latter of which has the added advantage of possessing protein, calcium and edible leaves. For over 7,000 years, such root vegetables have served as a primary source of carbohydrates for the region's inhabitants, and their leaves have served as fodder for domestic

---

(Lima: Oficina Nacional de Evaluación de Recursos Nacionales, 1984), 9; Iden, *Centros Poblados: Sexto Censo Nacional De Población, Primer Censo Nacional De Vivienda, 2 De Julio De 1961*. Vol. II. (Lima: Dirección Nacional de Estadística y Censos, 1966), 260; Carmen López Cisneros, *Huancavelica Ya No Es Tierra Del Mercurio* (Lima: Fundación Friedrich Ebert, 1983), 23.

18   Patiño Paúl Ortiz, *Huancavelica colonial*, 190–193; Guillermo Lohman Villena, *Las minas de Huancavelica en los siglos XVI y XVII* (Lima: Pontífica Universidad Católica del Perú, 1999), 52, 54; Rodrigo Cantos de Andrade, "Relación de la Villa Rica de Oropesa y minas de Guancavelica," in *Relaciones geográficas de Indias*. Vol. I. Marcos Jiménez de Espada, Ed. (Madrid: Ediciones Atlas, 1965), 304.

19   Espinoza Flores, 94, 99; Benjamin Waite, "Puya raimondii: Wonder of the Bolivian Andes," in *Journal of the Bromiliad Society*, Vol. II8, No. 5 (September–October, 1978), 200–202.

animals. Not only do potatoes prosper at higher elevations, many are easily preserved through freeze-drying, yielding the small, durable, *chuno*.[20]

Underscoring the importance of tubers in the regional diet is the fact that they occupy over one-quarter of the department's cultivated land, followed by corn, barley and wheat, fava beans and peas. Most inhabitants still plant according to the moon phase and utilize the traditional *chakitaklla*, a wooden device utilized for planting and harvesting tubers consisting of a pole about a meter long with a curved point and cross member upon which the foot is placed to penetrate the ground. Traditionally, tubers are sown in two phases, the first and smaller planting occurring in July and August, and the second, major, one in October. Harvesting begins around February, although the second crop may not be gathered until June.[21]

Huancavelica's farmers have traditionally rotated their crops, often replacing tubers with grains on a plot before it is left fallow and used for pasturing animals whose excrement fertilizes the soil. Overall, in a decade, a field will generally lie fallow for about six years, although with the introduction of new tuber cultivars this period has become shorter. Such improved cultivars not only mature more quickly, but they are more resistant to disease, have good market demand, and are suitable for making *chuño*. As land pressures increase, farmers are increasingly cultivating tubers above 3,600 meters. While this can be done successfully, crops at that altitude are more vulnerable to freezes and hail damage, and the land requires a longer fallow period.[22]

Just as the potato is a fixture of the Andean diet, so too have camelids played a central role in the human development, and customs, of the region since their domestication around 7,000 years ago. *Llamas, alpacas, guanacos,* and undomesticated *vicuñas,* are so prevalent in the higher elevations of the region that the department ranks as the third largest producer of such animals in Peru.[23] The two dozen fiber colors in the region come primarily from the long-haired *suri alpaca,* and the shaggy *ch'aku llama.* Another breed of llama, the short haired *k'ara,* serves primarily for transporting goods as opposed to

---

20   Espinoza Flores, 100; Stef de Haan, *Catálogo De Variedades De Papa Nativa De Huancavelica, Perú* (Lima: Centro Internacional de la Papa, 2006), 13–15, 21.

21   de Haan, *Catálogo De Variedades De Papa,* 11, 15, 14, 16–17.

22   Stef de Haan and Henry Juárez, "Land Use and Potato Genetic Resources in Huancavelica, Central Peru," in *Journal of Land Use Science,* Vol. v, No. 3 (2010), 179–180, 184–185, 192–193; Stef de haan, Jorge Nuñez, Merideth Bonierbale and Marc Ghislain, "Multilevel Agrobiodiversity and Conservation of Andean Potatoes in Central Peru: Species, Morphological, Genetic, and Spatial Diversity," in *Mountain Research and Development,* Vol. IIIo, No. 3 (2010), 223.

23   Cantos de Andrade, 307; Rivas Berrocal, 28; Espinoza Flores, 100, 105–106.

a source of wool. The best fibers, however, come from the diminutive vicuña, which has some of the softest, and most valuable, wool in the world. Because of this, in the 1970s, this shy creature was hunted almost to extinction, but in past decades has made a comeback. Since vicuñas cannot be domesticated, communities organize annual roundups, called a *chaccu*, in which the animals are corralled, sheared, and released, and their wool sold for the production of luxury clothing. High quality wool from alpacas, and the lesser quality wool of the llama, also provide the raw material for clothing, blankets, as well as straps and slings used for hunting.[24] Beyond keeping people warm, camelids keep them nourished. Llama and alpaca meat are good sources of protein, and that of the alpaca is cholesterol free. Rounding out their contributions, llamas and alpacas serve as pack animals, carrying up to about thirty-five kilograms each. They are also low maintenance creatures, with the alpha male shepherding the flock from the corral for a day of grazing, and leading them back in the evening.[25]

In addition to camelids, the region is home to Andean deer, known locally as *taruca (Hippocamelus antisensis)*, *vizcacha (lagidium peruanum)*, a long-tailed rodent which resembles a rabbit, and *cuy (cavia porcellus*, or guinea pig), all of which are protein sources. Other denizens of the region include the small, elusive and endangered Andean Mountain Cat (*Leopardus jacobita*), foxes, skunks and chinchillas, a relative of the vizcacha. There are also at least thirty-five species of birds, including the Puna Ibis, known locally as the *yanavico (plegadus ridgwayi)*, and the Andean goose, or *huallata (Chloephaga melanoptera)*.[26]

Conversing with the Cosmos

Indigenous Andeans often impute divine powers to geographical features, and this appears to have been the case with the cinnabar-rich Chacllatacana mount which lies above the city of Huancavelica. The term Huancavelica most likely derives from *huaca*, a religious object or place, and *villca*, a spiritually

---

24 Espinoza Flores, 106; G. Lichtenstein, G., R. Baldi, L. Villalba, D. Hoces, R. Baigún, and J. Laker, *Vicugna vicugna*. 2008 In The IUCN Red List of Threatened Species. Version 2014.3. <www.iucnredlist.org>. Downloaded on 21 February 2015; Raúl Zubilete, "En chakus logran más de una tonelada de fibra de vicuña." Correo del Sur. January 1, 2015. Accessed on January 1, 2015. http://diariocorreo.pe/ciudad/logran-mas-de-una-tonelada-de-fibra-554480/.

25 Government of Peru, *Inventario Y Evaluación De Los Recursos Naturales De La Zona Altoandina Del Perú: Reconocimiento, Departamento De Huancavelica*, Vol. II, 381, 386.

26 Espinoza Flores, 100, 105.

imbued geographical eminence, leading to the term to Huaca-villca. Another theory posits that it refers to a legend that the Huanta people, who lived in the region of Huanta and Churcampa, were said to have descended from a group known as the Willcas. They maintained their connection with their ancestors through worshipping a stone huaca in the Chacllatacana hill. Thus, the place became known as Huanca-willcas, or the shrine of the Willcas, which evolved into Huancavelica.[27] Despite the variations, all of these theories are based on a spiritual connection between residents and the Chacllatacana hill.

There are, in addition, some less plausible theories, such as one that the name derived from an extinct language that preceded Quechua and referred to a place where rock was formed by water. Also unlikely is that the place was named after a battle between an Inca military leader named Huamán, and a local chief, who lost the engagement. According to this view, Huamán named the place after him, and the nearby Villca hill. This seems improbable as it is rare that victors name places after their defeated rivals.[28]

Based on the foregoing, the word Huancavelica most likely reflects the deeply spiritual relationship between native Andeans and their environment; one based on complementarity, reciprocity, and trepidation of the harm nature can inflict. Rather than seeing themselves as divorced from their surroundings, indigenous Andean perceptions of the cosmos and natural environment reflect a view in which they are largely harmoniously integrated with it.[29] This reflects the Andean belief that the god Viracocha emerged from a mountain, created the cosmos and associated deities, and turned all that which preceded him to stone. He then populated the mountains, caves, rivers and land with people, from whom the indigenes descended. This myth underscores the polytheistic

---

27   Pedro de Ribera, and Antonio de Chaves y de Guevara. "Relación de la Ciudad de Gua-
      manga y sus terminos. Año de 1586," in *Relaciones geográficas de Indias*. Vol. 1. Marcos
      Jiménez de Espada, Ed. (Madrid: Ediciones Atlas, 1965), 182.
         Delgado de Castro, 13; Patiño Paúl Ortíz, 20; Salas Guevara, *Villa Rica de Oropesa*, (Peru:
      np, 1993), 32; Espinoza Flores, 24; Rivas Berrocal, 28; Government of Peru, *Almanaque De
      Huancavelica* (Lima: Instituto Nacional de Estadística e Informática, 2002), 18.
28   Salas Guevara, *Villa Rica de Oropesa*, 32; Murua, 549; Mariana Carolina de Belaunde and
      Ana Luisa Burga, *Huancavelica cuenta: temas de historia huancavelicana contados por sus
      protagonistas* (Lima: Instituto de Estudios Peruanos, 2005), 40; Espinoza Flores, 24; Del-
      gado de Castro, 13–14.
29   Daniela di Salvia, "La Pachamama En La Época Incaica y Post-Incaica: Una Visión Andina
      a Partir De Las Crónicas Peruanas Coloniales (Siglos XVI y XVII)/Pachamama in the Inca
      and Post-Inca Period: An Andean Vision from the Colonial Chronicles of Peru (16th and
      17th Centuries)," in *Revista Española de Antropología Americana* Vol. IV3, No.1 (2013), 96;
      Juan Ossio, "Cosmologies," in *International Social Science Journal*, Vol. IV9, No. 4 (1997), 549.

nature of indigenous Andean beliefs, although not all gods are seen as equal and instead fit into a spiritual hierarchy. For example deities such as the paternal *Inti*, the sun god, *Mama Quilla*, the moon goddess, and *Pachamama*, mother earth, enjoy elevated status due to their association with time, agriculture and fertility.[30]

One step down in this spiritual hierarchy are the innumerable *apus*, or mountain-dwelling gods who serve as fictional ancestors of allyus. Like all divinities, apus are believed to be generous yet unforgiving if their subjects fail to bestow offerings in the proper manner and at the appropriate time. Apus, who are generally male, are often venerated on "their" mountain, however they can also be invoked to visit a home or other place where offerings are made.[31]

Below them in divine rank are haucas, which have diverse putative powers, and are also often associated with geographical features such as lagoons, caves, mountains and rocks. Many huacas are viewed as familial ancestors who have been turned to stone, while others may be associated with a specific occupation. For example, many Peruvian miners venerate *Muqui*, a masculine and demanding deity who lives in mountains and determines their success, or even survival. Such beliefs highlight the intimate relationship between native Andeans and their natural surroundings, one which requires tribute to avoid upheaval in the form of droughts, earthquakes and other natural disasters.[32]

Andean cosmology is laden with dualistic concepts, such as the male sun and female moon, creation/destruction, harmony/disharmony, etc., with balance and harmony being maintained through reciprocity and the complementarity of opposites. Such ideas fostered the emergence of kings who were perceived as demigods and who served as intermediaries between the cosmos and their subjects to maintain cosmological and natural harmony. Spiritually,

---

30 Salvia, 90–95, 107; Luis Basto Girón, "Las mitas de Huamanga y Huancavelica," in *Perú Indígena* Vol. v, No. 13 (December, 1954), 10–12.

31 Nash and Williams, 458; Michael J. Sallnow, *Pilgrims of the Andes: Regional Cults in Cusco* (Washington: Smithsonian Institution Press, 1987), 127–128; Salvia, 107; Juan Ossio, "Cosmologies," in *International Social Science Journal*. Vol. IV9, No. 4 (1997), 556; Espinoza Gonzales, 67–68.

32 Heraclio Bonilla, "Religious Practices in the Andes and their Relevance to Political Struggle and Development," in *Mountain Research and Development* Vol. II6, No.4 (2006), 336, 339; Salvia, 95, 100; Kenneth Mills, "The Limits of Religious Coercion in Mid-Colonial Peru," in *Past and Present*, No. 145 (1994), 95; Iden, *Idolatry and It's Enemies: Colonial Andean Religion and Extirpation, 1640–1750* (Princeton: Princeton University Press, 1997), 75–76, 93, 210; Basto Girón, 11–14; Donna Nash and Patrick Williams, "Sighting the Apu: A GIS Analysis of Wari Imperialism and the Worship of Mountain Peaks," in *World Archaeology* Vol. II8, No. 3 (2006), 458; Mills, "The Limits of Religious Coercion," 95.

reciprocity is expressed through rituals, many of which and occur at transition-
al moments such as planting and harvesting. By making offerings such as coca,
food, alcohol, animals, textiles and harvest specimens, people seek to express
not only gratitude and reciprocity, but also recognize the complementarity between humans and their environment. Complementarity and the relationship
of dualism to totality is also expressed in the social organization of indigenous
communities, which are structured as two halves, or moieties.[33]

Beyond polytheism and dualism, a cyclical concept of time is also central to
Andean cosmology. Each epoch is said to last 1,000 years, and is dualistically di-
vided into two equal eras. Millennial epochs are begin and end with a *pachacuti*,
or massive cataclysmic upheaval, which is then followed by another prolonged
period of order. It was into the fifth epoch which the Francisco Pizarro stepped
in 1532, heralding the pachacuti which would upend indigenous life. Unlike
the Inca, who like their predecessors were polytheistic and assimilative, the
Spanish systematically endeavored to destroy indigenous religious beliefs and
practices. Their efforts in this regard were ultimately unsuccessful, and many
of the lifeways discussed above continue to this day.[34]

A Sacred Place

The rites through which indigenous Andeans expressed their veneration included the use of llimpi, or vermilion, which is derived from cinnabar. Not only
was llimpi blown on huacas as a sign of veneration, but it also was employed in
funeral rites. Sediment core studies from the Yanacocha and Negrilla Lagoons
reveal that cinnabar extraction in the region dates from 1400 BC, although it declined in the early Intermediate Period (200 BC to 500 AD). The primary source
of cinnabar in the region appears to have been the Chacllatacana mount, and
isotopic studies suggest that its cinnabar may have been traded with residents
of the distant north coast of Peru as far back as 1000 BC.[35]

The region of Huancavelica has thus long been imbued with spiritual and cosmological importance, and it was cinnabar which attracted both pre-Hispanic
peoples, and later Spanish colonists, to the region. Although lightly settled, people

---

33   Ossio, 558, 560–561; Salvia, 86, 98, 102, 107; Nash and Williams, 458; Sallnow, 3.

34   Ossio, 549, 555–556; Justin Jennings, "Inca Imperialism, Ritual Change, and Cosmological
      Continuity in the Cotahuasi Valley of Peru," in *Journal of Anthropological Research* Vol. V9,
      No. 4 (2003), 433, 443, 455; Nash and Willimas, 458; Salvia, 96,107; Bonilla, 336.

35   Patiño Paúl Ortíz, 21; Lohmann Villena, *Las minas de Huancavelica*, 13; Cooke, et al., "Over
      three millennia," 8833; Cooke et al., "Use and legacy of mercury," 4184.

of the Chavín culture (900 BC to 200 AD), established themselves at the site known as Chuncuimarca, in what is today the Asención neighborhood. The main regional Chavín administrative hub of Atalla, about sixteen kilometers from present-day Huancavelica, served as their primary collection and distribution point for cinnabar production. To obtain the cinnabar, pre-Hispanic workers would use deer antlers to extract the mineral. They would then pulverize it in a *maray*, or a mortar, often made of granite and approximately a meter in diameter, by means of a *quimbalete*, which served as a pestle and was about one-half the size of the mortar.[36]

Also in Huancavelica were settlers from the Paracas culture (800 BC to 100 BC), who resided in what is today the Santa Ana neighborhood, as indicated by the ruins of Segsachaka and Paturpampa. In addition to cinnabar, residents also mined obsidian, which was used for weapons and tools, and also dispatched wool and meat from camelids to their respective cultural centers. With the decline of the Chavín and Paracas cultures, came the rise of the Huarpa culture, (200 BC to 500 AD). Based in the region of Ayacucho, this group does not appear to have settled in what is today the city of Huancavelica.[37]

The rise of the Huari civilization (800 AD to 1200 AD), based just outside of present day Ayacucho, did lead to limited settlement near Huancavelica at Chimpamoqo. As with their predecessors, trash heaps reveal extensive consumption of vizcacha, llamas and alpacas, as well as stone and bone tools, worked copper and obsidian arrowheads. Cinnabar production increased under Huari domination, possibly in part to satisfy the demand by the Moché (100 AD to 700 AD) and Sicán (700 AD–1200 AD) civilizations in northern Peru, who made extensive use of it in burial rituals.[38] Following the decline of the

---

36   N.A., "Memoria sobre la mina de azogue de Huancavelica," in *Colección de memorias científicas, agrícolas é industriales publicadas en distintas épocas.* Vol. II. Mariano Eduardo de Rivero y Ustáriz, Ed.(Brussels: Imprenta de H. Goemaere, 1857), 85; Cobo, 150; Cantos de Andrade, 304; Fonseca, 347; Ramírez, "Descripción del reyno del Pirú," 321; Bargallo, *La minería y metalurgía,* 39–40; Crosnier, 36; Singewald, 518; Lohmann Villena, *Las minas de Huancavelica,* 11, 13; Patiño Paúl Ortíz, 20–21; José A. Hernández, "Antiguedad y actualidad de la Villa Rica de Oropesa," in *Peruanidad,* Vol. III, No. 12 (Lima, 1943), 942; Cooke, et al., 418–484; Cooke, et al. "Over three millennia," 8830; Ruiz Estrada, 30, 37; Bakewell, Miners of the Red Mountain, 15; ; Petersen, 38. For more on pre-Hispanic mining implements, see Petersen, 36–38.

37   Cantos de Andrade, 303; Ruiz Estrada, 30, 33–34, 37–38, 41; Government of Peu, *Almanaque de Huancavelica,* 16–17; R. Burger and R. Matos. "Atalla: A Center on the Periphery of the Chauvín Horizon," in *Latin American Antiquity* Vol. I3, No. 2 (2002), 164–167, 172–173; Brooks, *Peru Mercury Inventory,* 3.

38   Cooke, et al., "Over three millennia," 8833; Iden, "Use and legacy of mercury," 4181–4184.

Huari, a power vacuum in the region enabled the emergence of the Angaraes kingdom. This local group held a monopoly on llimpi production until they came under Inca rule in 1470, following their conquest by Capac Yupanqui, the brother of the ninth Inca ruler Pachacutec.[39]

In order to enforce their regional dominance, the Inca introduced settlers, and also relocated their combative new subjects to Jauja, Marcavelica and Llacsapallanca, the latter of which was the site of the present day city of Huancavelica. During the period of Inca rule, production of llimpi increased to unprecedented levels, as evidenced by sediment core samples from the Negrilla Lagoon. Like their forebears, the Inca utilized llimpi for rituals, war paint, and as a cosmetic for noble women, although they appear to be the first to excavate on the Santa Bárbara hill. Although Huancavelica was the principal source of cinnabar for the Inca, they also mined it in Chonta, Peru, and Cerro Colorado in present-day Bolivia. Underscoring the importance of the region to the Inca is the Temple of the Sun which they constructed in Incahuasi, outside of the present-day town of Huaytará. Situated in the upper reaches of the Pisco River valley, it was an administrative center on the *qhapaq ñan*, or the Inca road connecting Cuzco to Quito. This road also served as a conduit for colonists arriving from the regions of present-day Piura, Peru, and Quito.[40]

For millennia, Huancavelica's cinnabar deposits have drawn people to settle in the region. Both nomads and settlers utilized hat which the land had to offer, weaving ropes and baskets from ichu, and eating a variety of tubers, especially mashua, oca and olluco. Vizcacha, guinea pigs, partridge, ducks, Andean geese and deer served as sources of protein. So too did camelids, which also provided wool and served as pack animals.[41] Although the economic impact of Huancavelica is best documented in the colonial period, llimpi derived from

39   Ruiz Estrada, 42, 46; Eberth Serrudo, "El tampu real de Inkahuasi y la ocupación Inka en Huaytará," in *Inka Llaqta*. Vol. I (2010), 174; Salas Guevara, *Villa rica de Oropesa*, 17; Government of Peu, *Almanaque de Huancavelica*, 16–17.

40   Ruiz Estrada, 42, 46; Serrudo, 174; Salas Guevara, *Villa rica de Oropesa*, 17; Government of Peu, *Almanaque de Huancavelica*, 16–17; Cooke, et al., "Over three millennia," 8833; Iden, "Use and legacy of mercury," 4181–4184; Singewald, 518; Lohmann Villena, *Las minas de Huancavelica*, 11; Brown, "Worker's Health," 468–469. Serrudo, 173–174, 191; Franklin Pease, and Faura N. Domínguez, *Los Incas en la colonia: Estudios sobre los siglos XVI, XVII Y XVIIII en los Andes* (Lima: Museo National de Arqueología, Antropología e Historia del Péru, 2012), 157. For a list of additional deposits, see Petersen, 29.

41   Ruiz Estrada, 14, 18, 25, 33; Carrasco, 39; Oswaldo Rivas Berrocal, *Huancavelica: Bases Para El Desarrollo Económico Y Social Del Departamento Huancavelica* (Lima: Editorial Monterrico, 1989), 27; Government of Peru, *Almanaque De Huancavelica*, 15; Iden, *Censo Nacional De Población Y Ocupación, 1940*, Vol. VI, v.

cinnabar there was extensively traded in the region well before the arrival of the Spanish. It was used in rites by the Chavín and Paracas cultures, and, under Huari domination, was traded with the Moché and Sicán cultures in northern Peru where it was extensively utilized in elite burial rites. With the rise of the Inca, demand and production increased to new levels, as the region's red riches were widely exchanged. Llimpi was a spiritual currency, used to implore, thank and propitiate often fickle divinities.

Under Spanish rule, mercury refined in Huancavelica enabled unprecedented production of silver, and the emergence of the peso of eight as a global currency. With this, the impact of Huancavelica's riches expanded not just regionally, but globally. American silver traveled across continents, transformed economies and ultimately led to the rise of today's globalized economy. During this process, however, the environment, and people, of the region paid, and continue to pay, an extraordinarily high price.

CHAPTER 2

The Pachacuti: Colonization, Catastrophe and the Rise of Mercury

For the Andean population, the Spanish conquest was nothing less than a pachacuti, or cataclysm, that would transform population and settlement patterns in the region. While the concept of the pachacuti, and the cyclical view of time which it encompasses, helped indigenous people accept the changes wrought by the conquest, they were on an unprecedented scale. Waves of disease for which indigenes had no resistance engulfed the region, while survivors were forcibly relocated to newly established Spanish-style towns. Not only did these new population centers facilitate the further spread of disease, but they also ensured that native Andeans were available to labor for, and be spiritually indoctrinated by, their new rulers. The assault on their life ways was comprehensive, systematic and relentless, as people struggled to cope, and survive, in what was in many ways a "new world" for them as well.

"Without [Indians] There Will Be Neither Mercury Nor Silver...Nor Peru."[1]

While there were many causes of Indian depopulation in the first century of Spanish rule, without question the largest was disease. European diseases arrived in the Andes even before the Europeans, with the region suffering from an outbreak of measles or smallpox beginning in 1524 as traders carried viruses south from Panama. This paved the way for the Spanish conquest as the epidemic resulted in the death of the Inca Huayna Capac and members of the royal family. This sparked a civil war between rival heirs Huascar and Atahualpa, adding death through military combat to that of disease. It was into such conflict and chaos that Francisco Pizarro and his soldiers stepped, and which they so effectively exploited, in 1532.[2]

1　Damián de Jeria, Protector de los Naturales. Lima, January 10, 1604. AGI Lima, 34, No. 42.C. 6. "pues sin ellos ni habrá azogue ni plata... ni Perú."

2　Bonilla, "1492 y la población indígena," 106; Cook, *Demographic Collapse*, 68–69; Henry Dobyns, "An Outline of Andean Epidemic History to 1720," in *Bulletin of the History of Medicine*, Vol. III7, No. 6 (November–December, 1963), 494, 496; Nicolás Sánchez Albornoz,

A multitude of diseases would subsequently desolate the region. Smallpox, measles and influenza were the first to arrive, and were followed by typhus and plague in the 1580s. The impact of such maladies was magnified because the region had never been exposed to these diseases, and due to the facility with which they spread. In addition to a lack of resistance, factors such as population concentration, climate, and the degree and nature of interactions with non-locals affect the spread and mortality of disease. Generally, people at higher elevations in the Andes fared better than coastal provinces as the former lived in more dispersed settlements, and enjoyed a cool climate which inhibited the spread of mosquito-borne diseases such as yellow fever and malaria.[3]

Nevertheless, all were susceptible to diseases such as influenza and measles, which are transmitted through contact with infected bodily fluids or the air when people sneeze or cough and release droplets containing the viruses. In the case of measles, the virus can survive suspended in the air or on a surface for up to two hours after it is released by an infected person. When someone who has no resistance to measles breathes the infected air, or touches an infected surface and then rubs their eyes or contacts the inside of their mouth or nose, they are have over a ninety percent probability of contracting it.[4]

Once infected with the measles virus, the victim initially develops symptoms such as a fever, cough, watery and red eyes and a runny nose. These are the harbinger of a rash which breaks out on their forehead before spreading throughout the body. This phase usually lasts around three days and is accompanied by severe fever which then breaks, leading to the subsidence of the rash. In addition to the symptoms above, people afflicted by measles may contract pneumonia or encephalitis, both of which can cause death, especially in people who are otherwise malnourished or weak. The influenza virus, which arrived in Peru in 1558, shares similar symptoms with measles, minus the rash, and can also lead to pneumonia.[5]

The Population of Latin America: A History. Translated by W.A.R. Richardson (Berkeley: University of California Press, 1974), 61; Assadorian, "La crisis demográfica," 74; C.T. Smith, "Depopulation of the Central Andes in the 16th Century," in *Current Anthropology*, Vol. l1, Nos. 4–5 (1970), 458; Crosby, 51–52; Robins, *Mercury, Mining and Empire*, 15, 193.

3   David Noble Cook, *Born to Die: Disease and New World Conquest, 1492–1650* (Cambridge: Cambridge University Press, 1998), 206; Iden, *Demographic Collapse*, 207–208, 245, 247, 254–255; Robins, *Mercury, Mining and Empire*, 15.

4   Cook, *Born to Die*, 206; http://www.cdc.gov/measles/about/transmission.html; http://www.cdc.gov/flu/about/disease/spread.htm.

5   http://www.cdc.gov/measles/about/signs-symptoms.html; http://www.cdc.gov/measles/about/complications.html; http://www.cdc.gov/flu/about/disease/symptoms.htm; http://www.cdc.gov/flu/about/disease/spread.htm; Cook, *Demographic Collapse*, 67.

Smallpox, resulting from an infection of the *variola* virus, prospers in cool and arid climates and is primarily transmitted through physical contact with an infected person or their fluids. Like measles and influenza, however, it may also be spread through the air or when a person touches a tainted object. Following an incubation period of about two weeks, initial symptoms emerge. These include fatigue, high fever, aches and nausea, followed by oral sores which progress from red dots to lesions which break open and are extremely contagious. As this second phase passes and the eruptions heal, a rash emerges on the face and, over the course of about twenty-four hours, spreads throughout the body. Over the coming three days, the rash transforms into pustules which become inflamed and filled with liquid before turning hard and scabbing, a process which usually ends about two weeks after the initial rash. As the scabs begin to flake off, they leave scars in their place. People are contagious with the beginning of symptoms until all of the scabs are gone, however they are most contagious during the ten days after the rash appears. Those who perish from the disease usually succumb to internal hemorrhaging or fever.[6]

Unlike the above diseases, the bubonic plague is spread as a result of being bitten by fleas which had used rats, mice and other rodents as their host and became carriers of the *Yersinia pestis* bacterium. Humans are especially at risk during epizootics, or animal epidemics. As their hosts die, the fleas seek new ones such as humans, whom they then infect. Other means of infection are through contact with the fluids or tissue of an infected person or animal. Symptoms develop rapidly, and include a headache, fever, chills, fatigue, and seizures. The telltale sign of infection, however, is swollen lymph nodes, especially near the site of infection. Bubonic plague can progress to septicemic plague, which infects the blood, causes internal hemorrhaging, shock and death. This form may also be contracted directly through the unprotected handling of the flesh or fluids of infected animals. When the bacteria colonizes the lungs, pneumonic plague may develop, with symptoms including a cough, abdominal pains, bloody mucous, pneumonia, and shock which can lead to death. Like influenza, measles and smallpox, pneumonic plague may be transmitted through droplets released into the air when an infected person coughs or sneezes. These latter two forms of plague are the most fatal forms of the disease, however all can be deadly if they are left untreated.[7]

---

6 http://emergency.cdc.gov/agent/smallpox/overview/disease-facts.asp; David B. Martin, "The Cause of Death in Smallpox: An Examination of the Pathology Record," in *Military Medicine* Vol. 167, No. 7 (2002), 546, 550; Crosby, 53.

7 http://www.cdc.gov/plague/transmission/index.html;   http://www.cdc.gov/plague/symptoms/index.html,   http://www.nlm.nih.gov/medlineplus/ency/article/000596.htm;   http://

Like the plague, typhus is a bacterial disease spread by fleas, in addition to mites, lice and ticks. A person contracts murine typhus when they are infected with *Rickettsia typhi* bacteria, usually as a result of a flea bite. In contrast, epidemic typhus results from infection with the *Rickettsia prowazekii* bacteria, usually from lice. Beyond transmission through bites, people may contract typhus through breathing contaminated dust or when an infected parasite or their excrement is rubbed into an open sore.[8] People with murine typhus suffer from pain in the abdomen, back, joints and muscles, as well as a fever, nausea and a cough. Unlike smallpox, a rash first develops on the trunk of the body before spreading. Although more lethal, epidemic typhus has similar symptoms, although they are accompanied by chills, discomfort induced by light, and severe disorientation. Those who perish from the epidemic typhus usually die from pneumonia and kidney failure.[9]

Although bubonic plague is treatable with antibiotics, and there are currently vaccines for smallpox, measles and influenza, these diseases decimated America's native population, with pneumonia often the final cause of death. In what is today Peru, the population in 1520 is estimated to have been around 9,000,000 people. Over the next hundred years, largely due to disease, the population would experience a free fall of about ninety-three percent before bottoming out at around 600,000 survivors in 1620. In many cases, it seemed that hardly had one epidemic run its course than another began. For example, in the sixty-three years between 1572 and 1635, no fewer than nine epidemics of measles, smallpox and influenza swept through the land. After a lull of almost a century, yet another Andean epidemic of smallpox and influenza ravaged the region in 1719–20.[10]

---

www.cdc.gov/plague/symptoms/index.html;   http://www.cdc.gov/plague/symptoms/index.html; Cook, *Born to Die*, 206.

8   http://wwwnc.cdc.gov/travel/yellowbook/2014/chapter-3-infectious-diseases-related-to-travel/rickettsial-spotted-and-typhus-fevers-and-related-infections-anaplasmosis-and-ehrlichiosis, http://www.nlm.nih.gov/medlineplus/ency/article/001363.htm; Cook, *Born to Die*, 206.

9   http://www.nlm.nih.gov/medlineplus/ency/article/001363.htm.

10   http://www.cdc.gov/plague/resources/235098_Plaguefactsheet_508.pdf;   http://wwwnc.cdc.gov/travel/yellowbook/2014/chapter-3-infectious-diseases-related-to-travel/rickettsial-spotted-and-typhus-fevers-and-related-infections-anaplasmosis-and-ehrlichiosis; Bakewell, *Miners of the Red Mountain*, 109; Cole, *The Potosí Mlta*, 28; Ann Zulawski, *They Eat from Their Labor: Work and Social Change in Colonial Bolivia* (Pittsburgh: Pittsburgh University Press, 1995), 64–65; Cook, *Demographic Collapse*, 65, 67, 253; Dobyns, 499, 501, 509–512: Lohmann Villena, *Las minas de Huancavelica*, 253, 279; Robins, *Mercury, Mining and Empire*, 97.

Contemporaneous epidemics often resulted in higher mortality than had the diseases appeared individually. For example, while the mortality rate of smallpox in a population that had never been exposed to it is around thirty to fifty percent, and that of measles is between twenty-five and thirty percent, mortality rates can approach sixty percent when the two are combined. Beyond simultaneous diseases among a population lacking resistance to them, colonial mortality was exacerbated by malnutrition, general poor health, parasitic illnesses, colonial labor demands and suicide.[11] The effect of disease on the population continued to reverberate after an epidemic had run its course, as there were fewer people to reproduce, and hence subsequent generations were smaller. Despite the efforts of colonial clergy to stamp out *sirvancuy*, or trial marriage, and also concubinage in general, these practices helped to ensure the existence of future generations.[12] The cost in human lives was patent to contemporaries. One contemporary compared the Indians to a "candle, that gives light to all and itself is bit by bit consumed and used up."[13] Another, writing in 1623, warned that "without mines there will not be silver nor gold, and if there is no one to work them everything will be finished."[14]

Among the consequences of this human die-off was a decline of the viability of the vertical archipelago system as traditional exchange networks contracted or disappeared. Exacerbating this process was the fact that animals such as camelids, so important as pack animals and sources of food and wool, were also vulnerable to the plague. Moreover indigenous animals now competed for pasturage with animals introduced by the Spanish, such as cattle, oxen, horses, mules, donkeys, pigs, goats and sheep, all of which had few predators and reproduced quickly. Similarly, land which had been used to cultivate New World crops such as corn, cotton and potatoes increasingly was used to cultivate Old World products such as wheat, oats, barley, lettuce, cauliflower, melons, cabbage, radish, lettuce, olives and grapes.[15]

---

11   Balthasar Ramírez, 299; Josep Barnadas, *Charcas: origines históricos de una sociedad colonial* (La Paz: Centro de Investigación y Promoción del Campesinado, 1973), 321, 323, 326–332; Cobb, *Potosí y Huancavelica*, 53; Cook, *Demographic Collapse*, 59, 69–71, 253; Dobyns, 515; Sánchez Albornoz, *The Population of Latin America*, 51, 54, 60; Crosby, 45; Robins, *Mercury, Mining and Empire*, 14, 35, 193.

12   Cook, *Demographic Collapse*, 252–253; Nelson Manrique, *Colonialismo y pobreza campesina. Caylloma y el valle de Colca, siglos XVI–XX* (Lima: Desco, 1985), 138.

13   Doña Nieves, 278–279. "como la candela, que da lumbre a todos, y a ella se va poco a poco consumiendo y gastando."

14   Memorial del capitán Don Pedro Gutiérrez Calderón, 4. "no habiendo minas no habrá plata ni oro y si no hay quien las labre se acaba todo."

15   Bonilla, "1492 y la población indígena," 108, 110; Dobyns, 513; Crosby, 67–68, 74–77.

Over time, populations develop an immunity to such diseases. This can come about through three ways. One is "active" immunity, in which the body successfully fights the disease and develops immunity in the process. The second is "passive" immunity which is provided through the placenta or mother's milk. The final form is innate immunity, in which there is genetic resistance. While many European colonists had developed an innate resistance to diseases such as smallpox and measles, Indians in the Americas initially developed either an active or passive immunity to such illnesses. This process was aided by the fact that many diseases evolve to become less lethal, thus ensuring their own survival by not destroying their host.[16]

Although it was not spared, the highland region of Huancavelica offered some protection to the spread of these diseases thanks to cooler temperatures relative to the lowlands, and its relative isolation from large population centers. As a result, the region's demographic decline was closer to fifty to sixty percent. Nevertheless, Andeans there lacked immunity, and in addition to widespread epidemics also suffered from one of plague and diphtheria in 1613. Like the epidemic of 1719–20, illness and death reduced labor supply and affected mercury production. In July, 1759 Huancavelica, and its canine population, suffered another unidentified epidemic, characterized by fever, headache, difficulty hearing, oral bleeding, weakness and depression.[17]

Some authors, both colonial and contemporary, exaggerate the effects of mining in the demographic decline.[18] Throughout the Andes, disease was clearly the primary driver of the demographic implosion. The effects of epidemics

16 Cook, *Demographic Collapse*, 72–73.

17 Fernando Montesinos, *Anales del Perú*. Victor M. Maurtua, Ed., Vol. II (Madrid: Imprenta de Gabriel L y del Horno, 1906), 193, 196; Lohmann Villena, *Las minas de Huancavelica*, 146–147, 253; Fernández Alonso, 353–355; Cook, *Demographic Collapse*, 59, 114, 116, 143–144, 197, 252–254; Iden, *Born to Die*, 206; Barnadas, *Charcas*, 321; Bakewell, *Miners of the Red Mountain*, 109; Sánchez Albornoz, *The Population of Latin America*, 54–56; Assadorian, "La crisis demográfica," 73; Antonio de Ulloa, *Noticias americanas: Entretenimientos phisicos-históricos, sobre laAmérica Meridional y la Septentrianal Oriental* (Granada, Spain: Universidad de Granada, 1992), 202; Miguel Molina Martínez, *Antonio de Ulloa en Huancavelica* (Granada, Spain: University of Granada, 1995), 151; Robins, *Mercury, Mining and Empire*, 15, 137.

18 Armendaris, 132; Fray Buenaventura Salinas y Córdoba, *Memorial de las Historias del nuevo mundo Pirú* (Lima: Universidad Nacional Mayor de San Marcos, 1957), 305; Gabriel Fernández de Villabos,*Vaticinios de la pérdidia de las Indias y mano de relox* (Caracas: Instituto Panamericano de Geografía e Historia, 1949), 29; Nelson Manrique, 138; Eduardo Galeano, *Las venas abiertas de América Latina* (Mexico City: Ediciones Siglo Veintiuno, 1971), 49, 60–61.

were, however, exacerbated by the consequences of mine work and refining, which included mercury and arsenic poisoning, cave-ins, silicosis and respiratory infections from the climate. Beyond depopulation resulting from death, that of Huancavelica was compounded by people fleeing from mita-serving provinces. Others "disappeared" as priests, *corregidores*, or governors, *curacas*, or local leaders of Indian villages, and *hacendados*, or hacienda owners, concealed their charges for their own purposes.[19] In some cases, depopulation from these various causes was so severe that royal officials had no alternative than to face the facts and reduce mita levies.[20]

Viceroy Toledo and the New World Order

The population collapse was in full swing when Viceroy Francisco Toledo disembarked in Lima in 1569. Unfortunately for indigenous Andeans, his arrival heralded monumental changes in terms of how they lived and interacted with

---

19   Memorial del capitán Don Pedro Gutíerrez Calderón, 1–2; Emetherio Ramírez de Arellano, 1–2; Gastelumendi, 59; Luis Jerónimo Fernández de Cabrera, Conde de Chinchón, "Relación del estado en que el Conde de Chinchón deja el gobierno del Perú al Señor Virrey Marqués de Mancera," in *Colección de las memorias o relaciones que escribieron los virreys del Perú*, Vol. 11, Ricardo Beltrán y Rózpide, Ed. (Madrid: Imprenta del Asilo de Huérfanos del S.C. de Jesús, 1921), 71; Montesinos, 94, 205; Lizárraga, 102; Cook, *Demographic Collapse*, 205, 250; Sánchez Albornoz, *The Population of Latin America*, 53, 60; Crosby, 38–93; Tandeter, *Coercion and Market*, 25; John A. Fisher, *Silver Mines and Silver Miners in Colonial Peru, 1776–1824*. Monograph series No. 7 (Liverpool: Center for Latin American Studies, University of Liverpool, 1977), 10; Molina Martínez, *Antonio de Ulloa*, 81; Bakewell, *Miners of the Red Mountain*, 116; Wiedner, 369; Lohmann Villena, *Las minas de Huancavelica*, 441; Robins, *Mercury, Mining and Empire*, 35.
20   Memorial del capitán Don Pedro Gutíerrez Calderón, 1; Audiencia de Lima, "Relación que la Real Audiencia de Lima hace," 296; Francisco de Borja, Príncipe de Esquilache, "Relación que hace el Príncipe de Esquilache al Señor Marqués de Guadalcasar, sobre el estado en que deja las provincias del Perú," in *Memorias de los virreyes que han gobernado el Perú durante el tiempo del coloniaje español*. Vol. 1. M.A. Fuentes, Ed. (Lima: Librería Central de Felipe Bailly, 1859), 85; Melchor Liñan y Cisneros, "Relacion de Don Melchor de Liñan y Cisneros, dada al Señor Duque de la Palata, del tiempo de tres años y cuatro meses que gobernó, desde 1678 hasta 1681," in *Memorias de los virreyes que han gobernado el Perú durante el tiempo del coloniaje español*, Vol. 1, M.A. Fuentes, ed. (Lima: Librería Central de Felipe Bailly, 1859), 305; Navarra y Rocaful, 224; Luis de Velasco, "Relación del Sr. Virrey, D. Luis de Velasco, al Sr. Conde de Monterrey sobre el estado del Perú," in *Colección de las memorias o relaciones que escribieron los virreys del Perú*. Ricardo Beltrán y Rózpide, Ed. Vol. 1 (Madrid: Imprenta del Asilo de Huérfanos del S.C. de Jesús, 1921), 109; Salinas y Córdoba, 293.

their environment. In his mid-fifties, Toledo was the youngest son of the Count of Oropesa, and as a member of the military Order of Alcántara, had served in Europe and North Africa. His primary task as viceroy was to reinvigorate silver production, which had been considerably reduced in Potosí as a result of the exhaustion of the richest, easily smeltable, ores. Toledo was definitely the man for the job: forceful, unrelenting and comprehensive in his approach to problem solving. During his almost twelve year tenure, Toledo led military expeditions against the Arucanian and Chiriguano Indians in what is now Chile and southern Bolivia, respectively, and orchestrated the apprehension and subsequent beheading the last Inca, Túpac Amaru, in 1572. Most importantly, he engineered the shift from a smelting-based silver economy to one based on mercury amalgamation, which would have unprecedented human, environmental and global economic consequences.[21]

While mass death from epidemics led many Andeans to question the strength of their deities, their cosmological connections were further undermined by forced resettlement in new, Spanish-style villages. These new villages, called *reducciones*, were central to the overarching plan of the Spanish crown for the Americas to systematically deculturate the indigenous peoples and exploit their labor. For example, in the department of Huancavelica, the present-day villages of Chinchacocha, Hananguanca, Luringuanca and Atunjauja were all formed as reducciones. Among the effects of this forced resettlement program, however, was the severance of people's spiritual relationships with their birthplaces. Moreover, a person's place of residence was often a link in the chain of the vertical archipelago through which they contributed to, and obtained, a diverse array of products. By undermining the vertical archipelago system, people were increasingly forced into a cash economy as they frequently were compelled to purchase products which they had previously cultivated or traded.[22]

Exacerbating the situation, disease spread more quickly in the reducciones as the indigenous population was more concentrated and had increased interactions with non-Indians. Even once the worst of the population decline had

---

21   León Gómez Rivas, *El virrey del Perú Don Francisco de Toledo* (Madrid: Instituto Provincial de Investigaciones y estudios toledanos, 1994), 45–54, 49–50, 55–56; Arthur Zimmerman, *Francisco de Toledo, Fifth Viceroy of Peru, 1569–81* (Caldwell, Idaho: The Caxton Printers, Ltd., 1938), 45–46; 48–52, 54–55, 62, 66–67, 80, 84, 86–90, 109–117, 121–124; Robins, *Mercury, Mining and Empire*, 23–25.

22   Francisco López Caravantes, *Noticia general del Perú*. Marie Helmer, Ed. Vol 1 (Madrid: Ediciones Atlas, 1989), Vol. IV, 205–207; Bonilla, "1492," 108, 110; Zulaski, "Wages, Ore Sharing and Peasant Agriculture", 415; Assadorian, "La crisis demográfica," 86.

occurred and natives had acquired some resistance to European diseases in the eighteen century, life expectancy was abysmally short. This was especially the case among those who served the mita in Huancavelica or Potosí who often suffered from mercury intoxication and silicosis. In many places it often did not exceed twenty-five years. Given the forgoing, it is no surprise that Indians often sought to escape from the reducciones and return to their ancestral lands.[23]

A Hierarchy of Harm

Conquest, epidemics and forced resettlement were only the beginning of the indigenous pachacuti. Whether or not they lived in a reducción, they were subject to curacas, corregidores and clergy, all of whom had seemingly endless demands for their labor, and often their land. The Andean had gone from a person largely harmoniously and sustainably integrated into their physical and cosmological environment, to being little more than one more natural resource caught up in a tug of war by their colonial masters.

Curacas held tremendous power, and often considerable wealth, in indigenous Andean communities.[24] These positions were traditionally hereditary

---

23   Emetherio Ramírez de Arellano, 1–2; Don Juan de Dios Cavitas, indio principal del pueblo de Turco, provincia de Carangas, pidiendo se rebaje la contribución de mitayos del ayllo Ilanaca, parcialidad de Pumiri, en dicho pueblo, por falta de indios aptos. April 16, 1807, ABNB ALP Minas 130/8, 8, 15; Cook, *Demographic Collapse*, 79; Brian Evans, "Census Enumeration in Late Seventeenth-Century Alto Perú: The Numeración General of 1683–1684," in *Studies in Spanish Population History*. D.J. Robinson, Ed. (Boulder: Westview Publishers, 1981) 42; Thierry Saignes, "Las etnias de Charcas frente al sistema colonial (siglo XVI). Ausentismo y fugas en el debate sobre la mano de obra indígena (1595–1665)", in *Jahrbuch für Geschichte von Staat, Wirtschaft und Gesellschaft Lateinamerikas*, Vol. II1 (1984), 36–37; Robins, *Mercury, Mining and Empire*, 174.

24   Juan de Matienzo, "Carta a S.M. del licenciado Matienzo, con noticia de la residencia, que por encargo del Virrey, habia tomado al corregidor, alcaldes, oficiales y otros jueces de la Villa de Potosi. Describe el estado en que hallo las minas y lo que hizo para aumentar las rentas reales. Refiere el casamiento de Juan de Torres de Vera con la hija del adelantado Ortiz de Zarate, y aconseja que para el major gobierno de las Provincias del Tucumán y Paraguay, se junten en una sola, y se funden pueblos en el Tucumán y en el Rio de la Plata para el comercio directo con España." Potosí, December 23, 1577. In *La audiencia de Charcas: correspondencia de presidentes y oidores. Documents del Archivo de Indias*. Roberto Levillier, Ed. Vol. I (Madrid: NP, 1918), 21, 22–24; Diego García Sarmiento de Sotomayor, Conde de Salvatierra, "Relación del estado en que deja el gobierno de estos reinos el Conde de Salvatierra al Sr. Virrey Conde de Alba de Liste." In *Colección de*

and had survived Inca domination, which was generally assimilative and utilized indirect rule. Under the Spanish, as the fiscal guarantors of their subjects
curacas were liable for all of the exactions required by corregidor from their
communities. These included tribute, debts from the forced sale of merchandise, and providing the required number of draft laborers for mita service.
Curacas were also answerable to their local clergy, who had their own list of
labor and other demands. As hereditary local lords, curacas had a longstand-
ing and direct relationship with the communities over which they ruled. Over
time, however, they found it increasingly difficult, if not impossible, to meet
the plethora of financial and other obligations with which they were saddled.
While some went to court to contest the terms of their servitude, and suffered
reprisals, others simply renounced their position. Former curacas, however, retained a voice in community affairs as they generally served on the *hilacata*, a
council of local indigenous leaders. The decline of hereditary curacas led to the
rise of "interim" curacas, who were often *mestizos*, or of mixed indigenous and
Spanish descent. Appointed on an open-ended basis by the corregidor, interim
curacas were often not from the community they ruled, and hence had no organic links to it. Such appointments did not require higher-level bureaucratic
approvals, and interim curacas could just as easily be removed if they proved
to be non-compliant to the corregidor's directives. Complaints against curacas,
both interim and hereditary, commonly revolved around being overcharged
for tribute, incessant demands for unpaid services, access to community lands,
being coerced into serving the mita more than legally required, and floggings.[25]

---

las memorias o relaciones que escribieron los virreys del Perú. Ricardo Beltrán y Rózpide,
Ed. Vol. II. (Madrid: Imprenta del Asilo de Huérfanos del S.C. de Jesús, 1921), 87; José de
Acosta, *De procuranda indorum salute*, L. Pereña et al., Eds. (Madrid: Consejo Superior de
Investigaciones Cientificas, 1984), 505.

25   Expediente de las Diligencias practicadas en virtud de Real Provision por el Lisenciado
      Dn Custaquio Ferrera contra el Dor Dn Joseph de Barco y Oliva Cura del Beneficio de
      Cicacica, sobre la exaccion excesiva de Derechos Parrochiales y otros abusos a los indios.
      La Plata, April 20, 1769. ABNB, EC.1769.58, 39; No Title, La Plata, January 11, 1776. Archivo-
      Biblioteca Arquidiocesano "Monseñor Taborga,"(hereafter ABAS) Archivo Arzobispal,
      Clero, Tribunal Eclesiástico, 1–5; Testimonio de la representacion que hizo el senor Fiscal
      Promotor General por parte de don Lorenzo Apu Bedoya y Melchora de la Cruz Anaya,
      indios del pueblo de Toledo sobre visitadores eclesiásticos. December 16, 1753, La Plata.
      ABNB, EC.1753.144, 1; Sinclair Thomson, *We Alone Will Rule: Native Andean Politics in the
      Age of Insurgency* (Madison: University of Wisconsin Press, 2002), 10–12, 23, 30, 45, 70,
      78–79, 81, 104–105; David Cahill, "Curas and Social Conflict in the Doctrinas of Cuzco, 1780
      1814," in *Journal of Latin American Studies.* Vol. I6, No. 2 (November, 1984), 354–355; Sergio
      Serulnikov, "Customs and Rules: Bourbon Rationalizing Projects and Social Conflicts in
      Northern Potosí During the 1770s," in *Colonial Latin American Review.* Volume 8, No. 2,

Other dreaded tasks involved transporting goods in journeys which could last several months and for which they were not provided a subsistence wage.[26]

Although curacas exploited their charges for their own purposes and were subject to demands or local clergy, they primarily did the bidding of the corregidor. In addition to the corregidor's civil duties of collecting tribute, seeing to the timely dispatch of mitayos, and responding to executive orders, they served as a local magistrate. Although they were paid a modest salary for these responsibilities, it was revenue from the *repartimiento de mercancías*, or forced distribution of goods, which generated the majority of their income and also defined the value of a district. This system, so despised by the native population, emerged early in the colonial period, and expanded enormously

---

1999 248, 251; Jurgen Golte, *Repartos y rebeliones: Túpac Amaru y las contradicciones de la economia colonial.* (Lima: Instituto de estudios Peruanos, 1980), 18; Karen Spalding, *Huarochirí* (Stanford: Stanford University Press, 1984), 228; Los indios de las parcialidades de Laymes, Cullpas, Chayantacas, Carachas, Puracas y Sicoyas, 1–6, 13, 17, 20–36; Felipe Quispe, Isidro Quispe, Andrés Flores, Lucas Choque y Baltasar Aruquipa, indios del publo de Guarina, provincia de Omasuyos, sobre los avíos que su cacique omitio darles para venire a la mita de Potosí. La Plata, 1750–1775. ABNB, ALP Minas 28/2, 1; Capítulos puestos por los indios del pueblo de Calacoto, 3–4, 21; Don José Salas Ordóñez, visitador de la provincia de Sicasica, contra el cacique Felipe Alvarez. La Plata, September 8, 1765. ABNB, EC.1765.77, 1–21; Robins, *Mercury, Mining and Empire*, 152–154.

26   Autos criminales seguidos por Lúcas Alanza, indio, contra el Gobernador de Sacaca don Eduardo Ayaviri, por haberle inferido una paliza que casi le compromete su vida y varias injurias, además dicho Gobernador dispone mal de los Reales Tributos. La Plata, June 10, 1758. ABNB, EC.1758.22, 1–2; Antonio de Ayans, "Breve relación de los agravios que reciven los indios que ay desde cerca del Cuzco hasta Potosí, que es lo major y más rico del Perú, hecha por personas de muchas experiencia y Buena conciencia y esapasionadas de todo interés temporal y que solamente desean no sea Dios N.S. tan ofendido con tantos daños como los indios resciven en sus almas y haziendas y que la conciencia de Su Magestad se descargue mejor y sus Reales Rentas no sean defraudadas en nada sino que antes bayan siempre en continuación (1596)," in *Pareceres juridicos en asuntos de indias (1601–1718)*, Ruben Vargas Ugarte, Ed. (Lima: CIP, 1951), 39, 41; Alfonso Mesía Venegas, "Memorial del P. Alfonso Mesía Venegas, sobre la Cédula del servicio personal de los indios. 1603," in *Pareceres juridicos en asuntos de indias (1601–1718)*, Rubén Vargas Ugarte, Ed. (Lima: CIP, 1951), 104; Fernández de Castro, 158; na., "Pareceres de los Padres de la Compañía de Jesús de Potosí. 1610," in *Pareceres juridicos en asuntos de indias (1601–1718)*. Rubén Vargas Ugarte, Ed. (Lima: CIP, 1951), 122; Manuel Toledo y Leiva, "Parecer del P. Manuel Toledo y Leiva, Rector del Colegio de la Compañía de Jesús de Huancavelica sobre la Mita de Potosí, a petición del Sr. D.D. José Santiago Concha Oidor de Lima y Gobernador de Huancavelica, en virtud de R.C. expedida en Madrid el 6 de Diciembre de 1719." In *Pareceres juridicos en asuntos de indias (1601–1718)*, Rubén Vargas Ugarte, Ed. (Lima: CIP, 1951), 182; Robins, *Mercury, Mining and Empire*, 154.

when it was made legal in 1751 before it was banned, but never abolished, in 1784. Under the *reparto*, by which the system is also known, corregidors, would coerce Indians to purchase a variety of grossly overpriced and often poor quality products for which they had little or no use. These could range from silk stockings, velvet and blue hair powder to sick mules, paper and books.[27]

From the colonial perspective, this system had several benefits. Not only did it serve as the corregidor's primary revenue stream, but it also further forced Indians into a capitalist economy as reparto goods had to be paid for with silver. Instead of subsistence production or the bountiful barter of the vertical archipelago system, Indians now had no alternative but to seek remunerated work so that they could pay their imposed debts. Finally, in cases where European goods were distributed, their sale served as the final link in the mercantilist chain by providing a colonial market for such products.[28] Those who resisted purchases or did not pay on time usually were flogged, had their few possessions seized, and they and their family could be imprisoned or forced to labor to settle the debt. Beyond the abuses inherent to the reparto, corregidors also drafted Indians for personal service, pasturing, agricultural labor, transport of products, and the production of textiles or other products.[29]

The conquest brought not only disease, divorce from ancestral environments, alien overlords and unprecedented exploitation, but also a systemic

---

27   Patiño Paúl Ortíz, 245, 247.

28   Auto acordado por la Real Audiencia, sobre que los Corregidores no repartan géneros. La Plata, February 13, 1772. ABNB, EC.1772.59, 1; Garci Diez de San Miguel, *Visita hecha a la provincia de Chucuito por Garci Diez de San Miguel en el año de 1567*, Waldemar Espinosa Soriano, Ed. (Lima: Casa de la Cultura del Perú, 1964), 36; Scarlett O'Phelan Godoy, *Rebellions and Revolts in Eighteenth Century Peru and Upper Peru* (Colonge: Bohlau Verlag, 1985), 99–102, 109. For an official list of reparto goods and prices in Peru in 1784, see Escovedo, 44. For vivid descriptions of Corregidor treatment of Indians, see Jorge Juan and Antonio de Ulloa, *Discourse and Political Reflections of the Kingdoms of Peru* (Norman: University of Oklahoma Press, 1978), 69–101; Fisher, "Silver Production," 298; Robins, *Mercury, Mining and Empire*, 155–156.

29   Carta de Pedro Camargo. Lima, March 12, 1595. AGI, Lima, 35, #2, 2; Don Francisco José Ayra de Ariuto, cacique del pueblo de Pocoata, parcialidad de Hurinsaya, provincia de Chayanta, sobre los excesos del teniente de gobernador de los minerales de Titiri y Aullagas para con los indios. La Plata, 1691. ABNB, ALP Minas 147.4, 1–3; Carta de Juan Gutiérrez de León y Matías Díaz Rodo, oficiales reales en el asiento del Espíritu Santo, provincia de Carangas, a esta Real Audiencia: Los excesos cometidos por don Nicolás de Avalos y Ribera, corregidor de dicha provincia, contra Sebastían Cabezudo de Velasco, el principal ingeniero y aviador de aquélla, y contra los indios han hecho descaecer las labores de minas y disminuir los reales quintos. Carangas, September 16, 1664. ABNB, ALP Minas 96/5, 1–2; Robins, *Mercury, Mining and Empire*, 157.

effort to destroy indigenous religions and their physical representations. Although religious extirpation campaigns were enthusiastically supported by Viceroy Toledo, they were most vigorously carried out in the seventeenth century. This was especially the case under Archbishop Bartolomé Lobo Guerrero, from 1609 to 1622, and during the thirty years in which the more severe Pedro de Villagómez held office, from 1641 to 1671. These efforts did not simply employ preaching and catechism, but relied on torture, the identification and destruction of huacas, and the flogging, humiliation, banishment or incarceration of local religious leaders.[30] Extirpation campaigns were imbued with ceremony, as clergy and their retinues preached and admonished people to confess, repent and hand over both their huacas and indigenous religious leaders. After collecting the religious articles, clergy ordered them incinerated or otherwise destroyed and furtively disposed of to prevent their continued veneration. Indigenous communities generally had few alternatives other than to offer what was usually fleeting and superficial compliance. In rare cases, however, they refused to reveal anything regarding their huacas, and killed their persecutors. Overall, these campaigns were not simply motivated by religious zeal, but also by a desire for power and wealth. By eliminating, or reducing, the authority of local religious leaders, and in some cases the curacas who supported them, the power of the clergy was consequently enhanced. The hunt for huacas also could be profitable, as the shrines often contained gold or other valuables which the clergy appropriated.[31]

Despite the scope and duration of this assault, indigenous beliefs were remarkably resilient and adaptable, and largely survived the onslaught. The polytheistic nature of Andean religions fostered integrative and syncretic solutions to spiritual conflict.[32] In addition, even if a huaca was destroyed, if Indians could collect a shard of it, this was believed to retain all of its power and thus they continued to venerate it. In cases when there was no trace of a huaca,

---

30   Mills, "The Limits of Religious Coercion," 89–93, Iden, *Idolatry and Its Enemies*, 86, 170; Pierre Duviols, *La destrucción de las religions andinas (Conquista y colonia)* (Mexico: UNAM, 1977), 145, 252–253, 423–424, 426.

31   Mills, *Idolatry and Its Enemies*, 285; Iden, "The Limits of Religious Coercion," 92–93, 117; Luis Millones, "Religion and Power in the Andes: Idolatrous Curacas of the Central Sierra," in *Ethnohistory*. Vol. II6, No. 3 (Summer, 1979), 259–260; Nicholas Griffiths, *The Cross and the Serpent: Religious Repression and Resurgence in Colonial Peru* (Norman: University of Oklahoma Press, 1996), 147, 159–161, 245; Pierre Duviols, *La destrucción de las religions andinas (Conquista y colonia)* (Mexico: UNAM, 1977), 252–253, 257–258, 373–377, 398, 403, 407.

32   Mills. *Idolatry and It's Enemies*, 4–5, 74, 251, 256, 284–285; Iden, "The Limits of Religious Coercion," 99, 115; Duviols, 434; Griffiths, 26.

people often continued to worship its spirit. New huacas also emerged, such as those to protect communities from their new rulers. Finally, Indians increasingly relied on huacas which were geographical formations and which could not be destroyed. Such resilience and adaptability led some contemporaries to question the value of such campaigns. For example, in 1626 Lima Archbishop Gonzalo de Campo expressed his view that despite all their evangelical efforts, the Indians continued to practice their own religions much as they did before the conquest.[33]

Unlike the leaders of extirpation campaigns, local clergy were on the whole somewhat more tolerant of indigenous religious practices. Not only did this promote a degree of harmony in the community, but by turning a blind eye to such practices many clergy expected their parishioners to tolerate their own abuses or illicit activities.[34] Generally, the average Andean had more frequent interactions with his parish priest than his corregidor. One of the reasons for this was that clergy were the only non-natives permitted to live in Indian settlements. As a result, they had a disproportionate impact on the spiritual and temporal lives of Andeans. Like corregidors, clergy received a modest stipend, the appeal of their parish was determined by its revenue potential, and, like reparto goods, the prices they could legally charge for services was, in theory, limited by an *arancel*, or list of approved fees. Although some clergy owned shops in their parishes, their primary sources of revenue derived from saint's day celebrations, funerals and, to a lesser extent, marriages. Forced marriages were common in Peru as clergy arranged unions and endeavored to stamp out the age-old Andean practice of sirvancuy. Whether or not a union was voluntary, in addition to the cost of conducting the wedding, the newlyweds were required to present their priest with copious "gifts," which could involve dozens of animals and products. In addition, colonial clergy were known for not adhering to their vows of chastity, sometimes living with their lovers and children in the village. In other cases, they were accused of raping young female parishioners, whom they would then incarcerate. For those who turned out to be pregnant, marriage was the only way out of captivity.[35]

---

33   Duviols, 431–432, 436; Griffiths, 8; Mills, "The Limits of Religious Coercion," 95, 99, 110–111.
34   Duviols, 399, 400–401, 410, 437; Mills, "The Limits of Religious Coercion," 115.
35   Autos seguidos por los Curas de este Arzobispdo y demas sufraganeos, sre la suspencion de la Rl. Cedula que manda que los Ynds no paguen obvenciones. La Plata, September 25, 1760. ABNB, EC.1760.75, 29; Juan and Ulloa, 1–2,102–125, 280–316; "Representacion de la ciudad del Cusco, en el año de 1768, sobre excesos de corregidores y curas," in *Colección Documental de la Independencia del Perú.* Vol. I, Book II, "La Rebelión de Túpac Amaru: Antecedentes." Cárlos Daniel Válcarcel, Ed. (Lima: Comisión Nacional del Sesquicentenario de la Independencia del Perú, 1971), 36–37; Christine Hunefeldt, "Comunidad, curas

Much more profitable than marriages, however, were funerals, in which the clergyman often appropriated all of the possessions of the deceased. Often, last wishes were ignored as clergy held, and charged the bereaved for, more elaborate funerals than had been requested. Remarkably, even after death an Indian could continue to be a revenue source for their former priest, as he could schedule, and demand advance payment for, memorial masses for years ahead.[36]

Saint's day celebrations also figured among the primary revenue streams for clergy. Every town had a patron saint, and a *cofradía*, or religious brotherhood, to honor him. The local priest selected not only the financial sponsor of the event, but several people who would serve the priest in the coming year. Their duties were not limited to organizing the event, but included providing an array of services, such as pasturing animals, cooking, domestic services, supplying firewood and transporting goods. By the late eighteenth century, the number of cofradías, and people coerced to serve in them, had expanded, exacerbating the exploitative nature of the system. Despite its often involuntary

---

y comuneros hacia fines del periodo colonial: ovejas y pastores indomados en el Perú," in *Hisla*, No. 2, (1983), 5; Ward Stavig, "Living in Offense of Our Lord": "Indigenous Sexual Values and Marital Life in the Colonial Crucible," in *Hispanic American Historical Review*, Vol. VII5, No.4 (November, 1995), 602–603; Regina Harrison,"The Theology of Concupiscence: Spanish-Quechua Confessional Manuals in the Andes," in *Coded Encounters: Writing, Gender, and Ethnicity in Colonial Latin America*, Francisco Javier Cevallos-Candau, Ed. (Amherst: University of Massachusetts Press, 1994), 144; Cahill, "Curas and Social Conflict," 259, 262; Patiño Paúl Ortíz, 49, 54–55; Nicholas Robins, *Priest-Indian Conflict in Upper Peru: The Generation of Rebellion, 1750–1780* (Syracuse: Syracuse University Press, 2007), 31, 47–56, 84–88; Iden, *Mercury, Mining and Empire*, 160, 163.

36   Real Cedula. Aranjuez, May 2, 1752. ABNB, RC.571, 145; "Representación de la Ciudad del Cusco," 39–42; na, "Tensiones y pleitos en una doctrina de naturales," Lima, October 30, 1772, in Comite Arquidiocesano del bicentenario de Túpac Amaru. *Túpac Amaru y la Iglesia: Antología* (Lima: Edubanco, 1983, 87); na, "Informe." Azángaro, October 18, 1781, in *Colección de obras y documentos relativos a la historia antigua y moderna de las provincias de Río de la Plata*. Vol. IV. Pedro de Angelis, ed. (Buenos Aires:Librería Nacional de J. Lajouane & Cia, 1910), 420; Boleslao Lewin, *La rebelión de Túpac Amaru y los orígines de la independencia de Hispanoamérica* (Buenos Aires: Libreria Hachette, 1957), 231; Testimonio en fi6 de las cartas de los rebeldes, comisiones e informe que Diego Cristóbal Túpac-Amaru hizo al Exmo. Sr. Virrey de Lima, en respuesta del indulto Gral que libro. Peñas, November 15, 1781. ABNB, SGI.1781.248, 8–9; Juan and Ulloa, 109; O'Phelan Godoy, *Rebellions and Revolts*, 111, 114–116; Iden, "El norte y las revueltas anticlericales del siglo XVIII," in *Historia y Cultura*, No. 12. (Lima: Instituto Nacional de Cultura, 1979), 123; Hunefeldt, 6, 25; Robins, *Priest-Indian Conflict*, 56–63; Iden, *Mercury, Mining and Empire*, 161–162.

nature, service for the church in a cofradía or other capacity carried with it an exemption from the mita.[37]

Rounding out the constellation of coercion were hacendados, a group which could subsume corregidors, curacas, clergy, religious orders, miners, and other individuals. Haciendas replicated the colonial system in microcosm, with the owner having largely unbridled power on his estate, dispensing limited material benefits such as clothing and loans, and expecting lifelong loyalty and service in return. Those who labored on haciendas were generally bound to the owner by debt, and because debt was inheritable, many were born indentured. While haciendas in the region of Huancavelica produced foodstuffs and other products which could be sold in the mining town, they also served a vital role as a labor "reservoir" from which the hacendado could draw to rent out their subjects or apply them to work in their own mines or smelters. Although perpetual servitude was a steep price to pay, living on a hacienda offered some advantages. Not only was the owner responsible for the payment of his charges' tribute, but he protected them from the reparto and mita service.[38]

Natural Bonds and Unnatural Bondage: The Rise of Amalgamation

Prior to the rise of amalgamation, Indians produced silver through refining rich ores in smelters known as *guayras*. Synonymous with the Quechua word for wind, guayras were conical furnaces usually no more than two meters high, constructed of either ceramic or stone with multiple openings on the sides to allow the wind to serve as a bellows to produce the heat to smelt ore. They were simple, highly effective and, when made of ceramic, portable. This allowed refiners to place them on hilltops or other locations to take advantage of shifting wind patterns. The operator would load prepared ore in a container and charge the furnace with either wood, charcoal, animal excrement, or yareta.

---

37 Autos seg.s por los indios del Pueblo de Guagui contra el cura Dr. Don Pedro Márquez sobre varios ynjurias que les infirio. La Plata, May 18, 1754. ABNB, EC.1754.49, 1, 3, 16; "Tensiones y pleitos en una doctrina de naturales," 77; Juan and Ulloa, 104–105; "Representación de la ciudad del Cusco," 46–49; Olinda Celestino, and Albert Meyers, *Las Cofradías en el Perú: region central* (Frankfurt: Vervuert, 1981), 125, 134, 136; David Cahill, "Curas and Social Conflict," 244; Serulnikov, "Customs and Rules," 254; Patiño Paúl Ortíz, 46, 47; Robins, *Priest-Indian Conflict*, 7–93; Iden, *Mercury, Mining and Empire*, 160–161, 163.

38 Henri Favre, "La Industria Minera De Huancavelica En La Década De 1960," in *Boletín De Lima*. No. 161 (2010), 85; Iden, "Evolución y situación de las haciendas en la región de Huancavelica, Perú," in *Revista del Museo Nacional*. Vol. III3 (1964), 239–240; Iden, "La industria minera," 85; Robins, *Mercury, Mining and Empire*, 166.

Once fired, the ore released a lead and silver froth, which over a period of days was then skimmed off and repeatedly smelted, ultimately yielding silver. It was this technique which provided the initial impetus for colonial silver mining in the Andes, most notably in the city of Potosí.[39]

Despite their simplicity and minimal production and operation costs, the guayra was most useful when smelting the richest, but least plentiful, ores. The result was a sharp rise, and then fall, of silver production in many mining towns, as the best, and most accessible ores, were soon depleted. Tons of unprocessed, second quality, ores literally piled up in silver mining towns throughout the region, and could only be profitably processed with the introduction of the mercury amalgamation system. Quicksilver will naturally bind with silver and gold, and the technique dates from at least Roman times. In the 1550s, the Sevillian Bartolomé de la Medina experimented with the process in Pachuca in New Spain, having learned of it in Europe. After several setbacks, by 1554 he had honed the technique and it was soon adopted throughout New Spain.[40]

---

39   Juan López de Velasco, *Geografía y descripción universal de las Indias* (Madrid: Establecimiento Tipográfico de Fortanet, 1894), 504; Robert West, "Aboriginal Metallurgy and Metalworking in Spanish America: A Brief Overview," in *Mines of Silver and Gold in the Americas*, Peter Bakewell, ed. (Brookfield, VT: Variorum, 1997), 50; Reginaldo de Lizárraga, *Descripción del Perú, Tucumán, Río de la Plata y Chile.* (Buenos Aires: Union Académique Internationale/ Academia Nacional de la Historia, 1999), 187; Francisco López Caravantes, *Noticia general del Perú.* Vol. II. Marie Helmer, Ed. Madrid: Ediciones Atlas, 1989, 138; 138; Luis Capoche, *Relación general de la Villa Imperial de Potosí.* Lewis Hanke, Ed. (Madrid: Ediciones Atlas, 1959), 110–111; José de Acosta, *Historia natural y moral de las Indias.* José Alcina Franch, Ed. (Madrid: Historia 16, 1987), 238; Ramírez, 349–350; Cobo, 144; José Eusebio Llano Zapata, *Memorias histórico, físicas, critico, apologéticas de la América Meridonal.* Ricardo Ramírez Castañeda, et al., Eds. (Lima: Instituto Francés de Estudios Andinos, 2005), 165; Bakewell, *Miners of the Red Mountain*, 15–16; Cobb, *Potosí y Huancavelica*, 84–87; Tandeter, *Coercion and Market*, 75; Barnadas, *Charcas*, 363–364; Bargallo, *La minería y metalurgía*, 40–41, 97; Purser, 24; Julio Sánchez Gómez, "La ténica en la producción de metales monedables en España y en América, 1500–1650," in *La savia del imperio: tres estudios de economia colonial*, Julio Sánchez Gómez, Guillermo Mira Delli-Zotti and Rafael Dobado, Eds. (Salamanca: Ediciones Universidad Salamanca, 1997), 49; Lope García de Castro. "Carta a S.M. del Licenciado Castro acerca de las minas y del trabajo de los indios; la sucesión de encomiendas; guerra contra los indios en Chile; fundación de un monasterio de monjas; y otros asuntos de menor importancia," in *Gobernantes del Perú: Cartas y Papeles siglo XVI.* Vol. III. Roberto Levillier, Ed. (Madrid: Sucesores de Rivadeneyra, 1921), 288; Pedro Cieza de León, *Crónica del Perú* (Lima: Pontífica Universidad Católica del Perú, 1984), 291–292; Matienzo, 133; Robins, *Mercury, Mining and Empire*, 18.

40   Lohmann Villena, *Las minas de Huancavelica*, 15; Jerome Nriagu, "Mercury pollution from the past mining of gold and silver in the Americas," in *The Science of the Total Environment*, Vol. 149 (1994),168–169; Sánchez Gómez, 84; Fisher, *Silver Mines and Silver Miners*, 3; Alan Probert, "Bartolomé de Medina: The Patio Process and the Sixteenth Century Silver

As practiced in the Andes, the mercury amalgamation process involved grinding the silver-bearing ore to a fine, almost dusty, consistency, and mixing it in two and one quarter metric ton batches in a container, or *cajón*, with water. Depending on the refiner's estimate of how much silver the ore contained, he would add about 220 kilograms of salt, forty five kilograms of copper pyrites or magistral, three kilograms of lime, and about two kilograms of iron, and smaller quantities of tin, lead and other elements. Refiners then added between ninety and 220 kilograms of mercury, depending on the class of argentiferous ore, sprinkling it into the mixture by squeezing it through a cloth. The desired consistency of the mixture was that of a pasty liquid, for if it was too runny the mercury would fall to the bottom, and if it was too thick it would not dissipate. This *torta*, as the mixture was called, was then transferred to a flat, walled area around twelve by three meters, leading to it being called the "patio process" by which it is also known. Ideally, the "patio," which was also called a a *buitrón*, would be built atop a chamber which would be charged with combustible material to heat the mass from below, as heat expedites the process. As many mining areas were located in desolate, un-forested areas, refiners commonly utilized the "cold" process instead. Although the cold amalgamation process could take up to a month, it did not involve the costs of purchasing and transporting firewood, or constructing the patio above a fire chamber.

Once spread in the buitrón, Indian workers known as *repasaris* would wade into the waist deep torta and "tread," or walk in, it while also mixing it with shovels to facilitate the binding of the silver and mercury.[41] Once the refiner determined that the amalgamation process had finished, Indians would transfer the torta to a series of connected, water-filled troughs where it was "washed" over the period of a day or two to allow the heavier amalgam

Crisis," in *Mines of Silver and Gold in the Americas*, Peter Bakewell, Ed. (Brookfield, VT: Variorum, 1997), 110–111; Peter Bakewell, "Registered Silver Production in the Potosi District, 1550–1735," in *Jahrbuch fur Geschichte von Staat, Wirtschaft und Gesellschaft Lateinamerikas*, Vol 12. (1975), 81;Robins, *Mercury, Mining and Empire*, 21.

41   Juan Alcalá y Amurrio, Directorio del Beneficio del Azogue, en los Metales de Plata. Documentos que se dan en sus Reglas. Oruro, 1781, ABNB, Ruck 80, 12–14, 21, 26, 52–53; Capoche, 123; Cobo, Vol. I, 147; Salinas y Córdoba, 264; Ulloa, *Noticias americanas*, 256; Caravantes, Vol. IV, 104–105; Pedro Vicente Cañete y Domínguez, *Guía histórica, geográfica, física, política, civil y legal del gobierno e intendencia de la provincia de Potosí* (Potosí, Bolivia: Editorial Potosi, 1952), 50, 65; Haenke, 126–130; Llano Zapata, 169; Cobo, Vol. I, 147; Bargallo, *La minería y metalurgía*, 110–112, 192, 291; Bakewell, *Miners of the Red Mountain*, 21–22, 148–149; John Fisher, "Silver Production in the Viceroyalty of Peru, 1776–1824," in *Mines of Silver and Gold in the Americas*. Peter Bakewell, ed. (Brookfield, VT: Variorum, 1997), 289; Iden, *Silver Mines and Miners*, 54; Cobb, *Potosí y Huancavelica*, 95; Brading and Cross, 554; Purser, 48; Robins, *Mercury, Mining and Empire*, 10–11, 86.

to settle at the bottom. This amalgam, or *pella*, would then be placed into cylindrical, cheese-cloth like fabric forms and compressed to remove, and reuse, as much mercury as possible.[42] What remained, which was about seventeen percent silver and eighty three percent mercury, was then loaded in about forty-five kilogram increments into conical ceramic molds, known as *piñas* due to their resemblance to pineapples. These were then placed on metal stakes, covered with a matching mold, and fired over a period of about eight hours. This vaporized the quicksilver, some of which condensed in the covering mold and could also be reused. The result was about seven kilograms of honeycombed silver in the piña, which was then usually transformed into silver bars, coin or decorative objects.[43]

This system was only slowly adopted in the Andes, reflecting both the initial abundance of rich ores and the reluctance or financial inability of miners to make the large investment necessary to construct the mill complexes. By the 1560s in Potosí, however, ore quality had declined considerably, and along with it so had the viability of the guayra system. Moreover, the cost of obtaining processable ore in the famed Cerro Rico which dominates the city was increasing as miners owner had to bore deeper shafts which were often prone to flooding. As ore quality declined in Potosí, so too did the city's population and the local and regional economy.[44]

42   Alcalá y Amurrio, 20, 53; Cobo, Vol. I, 147–148; Bartolomé Arzáns de Orsúa y Vela, *Historia de la Villa Imperial de Potosí*. Lewis Hanke and Gunnar Mendoza, Eds. Vol. I, (Providence, RI: Brown University Press, 1965), 170; Fisher, *Silver Mines and Silver Miners*, 54; Renate Pieper, "Innovaciones tecnológicas y problemas del medio ambiente en la minería novohispana (siglos XVI al XVIII)," in *Europa e Iberoamerica: Cinco siglos de Intercambios*, Vol. II. María Justina Sarabia Viejo, et al., Eds. (Seville, Spain: Asociación de Historiadores Latinoamericanistas Europeos/ Consejeria de Cultura y Medio Ambiente, 1992), 357–358; Buechler, "Technical Aid," 53–54; Caravantes, Vol. IV, 105; Cobo, Vol. I, 148; Cañete y Domínguez, 65; Bakewell, *Miners of the Red Mountain*, 22; Robins, *Mercury, Mining and Empire*, 87–88.

43   Don Rodrigo de Mendoza y Manrique, 2; Cañete y Domínguez, 65; Capoche, 124; Arzáns, Vol. I, 171; Cobo, Vol. I, 148–149; Purser, 50; Cobb, *Potosí y Huancavelica*, 90; Bakewell, *Miners of the Red Mountain*, 22; Bargallo, *La amalgamación*, 213; Robins, *Mercury, Mining and Empire*, 88.

44   Lizárraga, 186–189; Capoche, 135; N.A. "Descripción del la villa y minas de Potosí. Año de 1603," 375; Llano Zapata, 165; Juan de Mendoza y Luna, Marqués de Montesclaros, "Relación del estado del gobierno de estos reinos que hace el excmo. Señor Don Juan de Mendoza y Luna, Marqués de Montesclaros, al excmo. Señor Príncipe de Esquilache su successor," in *Memorias de los virreyes que han gobernado el Perú durante el tiempo del coloniaje español*, Vol. I, M.A. Fuentes, Ed. (Lima: Librería Central de Felipe Bailly, 1859),

Although Andean miners were hesitant to make the transition to the amalgamation system, King Charles I (1500–1558) was not. He recognized, however, that while the mercury mine in Almadén, Spain, could provision New Spain and Central America, it could not satisfy the needs of the Viceroyalty of Peru once the system was adopted there.[45] In 1555, just one year after Medina had demonstrated the viability of the system in Pachuca, Charles I ordered his new viceroy of Peru, Andrés Hurtado de Mendoza, Marqués de Cañete, to encourage prospecting for cinnabar and the use of the system in the region. Although a small cinnabar lode was modest deposit was identified by Gil Ramírez near Cuenca, present day Ecuador, in 1558, it was by no means sufficient to supply the viceroyalty. The next year, two other prospectors, the Portuguese Enrique Garcés and Pedro Contreras, encountered two women selling llimpi in the market in Lima's main plaza. Upon questioning, the vendors revealed that it had come from Tomac and Guacoya, villages near Guamanga, present day Ayacucho. The two prospectors began to seek the source, and not only unearthed cinnabar, probably in Tomac, but refined mercury from it.[46]

Although the partnership between Garcés and Contreras dissolved soon afterwards, their work had come to the attention of Viceroy Cañete. The viceroy subsequently dispatched Garcés to New Spain to study the amalgamation system. In 1559, upon his return to Peru, Garcés became an advocate for the system and was rewarded with a monopoly on all mercury production in Guamanga, Huánuco and Lima, subject to paying 25% of his production to the

---

37; Cobb, *Potosí y Huancavelica*, 88; Bakewell, *Miners of the Red Mountain*, 8; Brading and Cross, 547–548, 555, 574; Robins, *Mercury, Mining and Empire*, 20.

45   Montesinos, Vol. II, 21; José Antonio Manso de Velasco, Conde de Superunda, "Relacion que escribe el conde de Superunda, Virrey el Perú, de los principales sucesos de su gobierno, de Real Orden de S.M. comuncada por el Excmo. Sr. Marqués de la Ensenada, su secretario del Despacho universal, con fecha 23 de Agosto de 1751, y comprehende los años desde 9 de Julio de 1745 hasta fin del mismo mes en el de 1756," in *Memorias de los virreyes que han gobernado el Perú durante el tiempo del coloniaje español*, Vol. IV, M.A. Fuentes, Ed. (Lima: Librería Central de Felipe Bailly, 1859), 165–172; Antonio Matilla Tascón, *Historia de las minas de Almadén*. Vol II. (Madrid: Instituto de Estudios Fiscales/ Minas de Almadén y Arrayanes, 1987), 113, 117; Bakewell, "Technological Change," 84; Brown, "La crisis financiera," 354, 362; Patiño Paúl Ortíz, 10; Probert, 97–98; Sánchez Gómez, 88–89; Bargallo *La minería y metalurgía*, 119–120; Iden, *La almagacion*, 90; Lohmann Villena, *Las minas de Huancavelica*, 57; Cobb, *Potosi y Huancavelica*, 35; Robins, *Mercury, Mining and Empire*, 9.

46   Lohmann Villena, *Las minas de Huancavelica*, 16; Cobb, *Potosí y Huancavelica*, 32, 35; Patiño Paúl Ortíz, 19, 166–167, 176–177, 179–180, 196; Bargallo, *La minería y metalurgia en la América*, 77; Cook, *Demographic Collapse*, 203; Goldwater, 46; Fisher, *Silver Mines and Silver Miners*, 3; Robins, *Mercury, Mining and Empire*, 21–22.

crown. The following year, with mercury extracted from Palca, in the present-day department of Huancavelica, Garcés refined ninety kilograms of silver, the first known case of mercury amalgamation in Peru, and one which eliminated any doubts as to the viability of the technique. Unfortunately for Garcés, his monopoly was rescinded in 1561 with the arrival of Viceroy Diego López de Zúñiga, Conde de Nieva. Despite this change in fortune, Garcés would serve as a *regidor*, or city councilman, in Huancavelica in the 1580s, having demonstrated that the region contained cinnabar deposits which could, at least partly, fuel a shift to an amalgamation-based system. A much larger source of quicksilver, however, would be necessary for it to be sustainable.[47]

Despite the demonstrated efficacy of the system, the cost of constructing refining mills, and miner's concerns about receiving a reliable supply of mercury, continued to inhibit the wider adoption of the system in the Andes. Two factors would, however, change the nature of this equation: the discovery of the rich and abundant cinnabar deposits in the Santa Bárbara mount outside of Huancavelica, and the reforms undertaken by Viceroy Toledo to transition to the new system.

To better understand and see the amalgamation process first hand, Toledo summoned Pedro Fernández de Velasco. Like Garcés, Fernández had studied the technique in New Spain. Convinced that the system could be effectively introduced to the region, Viceroy Toledo dispatched Fernández to Potosí to urge its adoption there. Beyond educating refiners in the new technique, Toledo understood that its widespread adoption required five components: hydraulic mills to process ore, water to power them, miners who were willing to invest in them, thousands of indigenous laborers, and mercury. To begin this undertaking, Toledo journeyed to Potosí, arriving there in late December, 1572.[48]

47　　Antonio de la Calancha, *Crónica moralizada del orden de San Augustín en el Perú, con sucesos egemplares en esta monarquía*, Vol. v (Lima: Universidad Nacional Mayor de San Marcos, 1974), 1679; Montesinos, Vol. II, 18–19; Bakewell, "Technological Change in Potosi," 78; Guillermo Lohmann Villena, "Enrique Garcés, descubridor del mercurio en el Perú, poeta y arbitrista," in *Studia*, Vol. II7–28 (August-December 1969), 15–20, 23–26, 41–42, 61; Iden, *Las minas de Huancavelica*, 19; Goldwater, 46; Cobb, *Potosí y Huancavelica*, 32; Lohmann Villena, *Las minas de Huancavelica*, 19, 117; Patiño Paúl Ortíz, 274; Cook, *Demographic Collapse*, 203; Robins, *Mercury, Mining and Empire*, 22.

48　　Francisco de Toledo, "Memorial que D. Francisco de Toledo dió al Rey nuestro señor, del estado en que dejó las cosas del Perú, después de haber sido en él Virrey y Capitán General trece años, que comenzaron en 1569," in *Colección de las memorias o relaciones que escribieron los virreys del Perú*, Vol. I, Ricardo Beltrán y Rózpide, Ed. (Madrid: Imprenta del Asilo de Huérfanos del S.C. de Jesús, 1921), 101; Acosta, *Historia natural,* 244; na, "Historia de la mina de Huancavelica," 65; Fisher, *Silver Mines and Silver Miners*, 3; Peter Bakewell, "Technological Change in Potosi: The Silver Boom of the 1570s," in *Mines of Silver and Gold*

At this time, about forty mills were already using hydraulic power to crush ore on the Tarapayá River outside of town. Although these stamp mills operated year round, their distance from the Cerro Rico entailed considerable transportation costs for the ore, and risk of theft in the process. As a result, most refiners utilized human or animal power to crush ore. Although these mills were much less costly to construct and operate, they paled in terms of efficiency when compared to those powered by water: a hydraulic mill was five times more efficient than one which used animal power and ten times more productive than those based on human labor.[49]

To overcome the challenge of providing a consistent water supply to mills whose construction was planned closer to the Cerro Rico, Toledo looked no farther than the hills which surround Potosí. Utilizing the natural forms of the land, he ordered eighteen interlocking reservoirs to be built almost 600 meters above the city, which would be filled by rain, snowmelt and spring water to provide a year round water source to energize the mills. This engineering marvel, replete with sluice gates and ten meter thick containment walls, was complemented by another one: a seventeen kilometer long, eight meter wide, stone-lined canal, called the *ribera*, which would carry the water to the mills. Work on the reservoirs began in 1574 when Toledo assigned 20,000 indigenous laborers to the task, and by 1580 water from them would link over 100 mills on the ribera. The system would subsequently be expanded, and by 1621 it consisted of thirty-two lagoons.[50]

Having addressed the issues of water supply and delivery, Toledo still had to overcome the reluctance of the mining *gremio*, or guild, to take on the expense, and risk, of constructing stamp mill complexes. He accomplished this through two ways. One was the development and implementation of a mining code, which was led by Toledo's seasoned advisors Juan de Matienzo and Juan Polo

in the Americas. Peter Bakewell, Ed. (Brookfield, VT: Variorum, 1997), 76, 81–82; Patiño Paúl Ortíz, 167; Lohmann Villena, *Las Minas de* Huancavelica, 59; Iden, "Enrique Garcés," 26; Robins, *Mercury, Mining and Empire,* 26.

49   Arzáns, Vol. I, 146; Montesinos, Vol. II, 61; Mendoza y Luna, "Relación del estado del Gobierno", 38; Murua, 570; Capoche, 118; Enrique Tandeter, "Forced and Free Labor in Late Colonial Potosí," in *Past and Present,* Vol. IX3 (1981), 110; Bakewell, "Technological Change in Potosi," 89, 92; Robins, *Mercury, Mining and Empire,* 28.

50   Cañete y Domínguez, 90; Capoche, 117; Arzáns, Vol. I, 157, 169; Rose M. Buechler, *The Mining Society of Potosi. 1776–1810* (Syracuse: Syracuse University Department of Geography, 1981), 39; Cobb, *Potosí y Huancavelica,* 91; Willaim E. Rudolph, "The Lakes of Potosi," in *The Geographical Review,* Vol. II6, No. 4 (October, 1936), 531–532, 535–536, 543;, 166; Bakewell, *Miners of the Red Mountain,* 13; Bargallo, *La amalgamación,* 206; Robins, *Mercury, Mining and Empire,* 27.

de Ondegardo. When the regulations were announced in 1574, they included provisions for debt protection for silver and cinnabar miners through exempting them from productive asset seizure, if they became bankrupt, so long as the crown was not the creditor. Within a decade, the code had become the law of the land throughout the Americas.[51]

Toledo's final inducement for miners to embrace the system was by providing them with an Indian draft labor allotment which depended on the length of their mining claim or the how many stamps they had in their mill. This system, known as the mita, had earlier been used by the Inca to mobilize labor for construction, infrastructural and other projects. Under the Spanish, it applied to rural able-bodied tributary male Indians, or those between eighteen and fifty years old, who were *originarios*, or who lived in their town of origin. In theory, about one-seventh of this population was to be called up each year, meaning that an Indian would only serve as a mitayo for one year out of every seven They were assigned to work on projects or provide services broadly deemed for the "public good," which did not preclude private benefit. Service could include labor in mining and refining, construction and infrastructure projects, work in obrajes and agriculture and pastoral work. There were some exemptions, such as for curacas and their eldest son, members of the hilacata members, those fulfilling a service to his parish priest, and the infirm.

As applied to mining in Potosí, 13,500 Indians from sixteen provinces in the region were assigned to a year of toil in mining and refining tasks. In Huancavelica, 3,289 men would be called up from twelve surrounding provinces to serve for two months, the shorter term in recognition of the toxicity of their task. Prior to the late 1600s when mercury refining was conducted year round, the Huancavelica levy was reduced by half during the wet season from

---

51   Expediente seguido por don Roque de Reinalte, sobre la prisión a que le han reducido por cierta deuda no obstante de su calidad de minero en Potosí. La Plata, August 14–19, 1697. ABNB, ALP Minas 19/10, 1, 5; Alberto Crespo Rojas, *La Guerra entre vicuñas y vasconga-dos (Potosí, 1622–1625)*. (Lima: Tipografía Peruana, 1956), 41–42, Zimmerman, 177, 183; 185; Patiño Paúl Ortíz, 105; Cobb, *Potosí and Huancavelica*, 55; Robins, *Mercury, Mining and Empire*, 27. The regulations are published as Francisco Toledo, Toledo, "Ordenanzas que el Señor Viso Rey Don Francisco Toledo hizo para el buen gobierno de estos Reynos del Perú y Repúblicas de él." In *Relaciones de los Virreyes y Audiencias que han gobernado el Perú*. Vol. I. Sebastian Lorente, Ed. (Lima: Imprenta del Estado, 1867), 267–348. For a summary of the code, see Zimmerman, 178–85 and W. Jakob, "Sumario de las ordenanzas mineras del Perú," in *Revista del Instituto de Historia del derecho Ricardo Levene*, Vol. III (Buenos Aires, 1972), 273–288.

January to May. Although Toledo codified the mita system, forced labor was nothing new in both Huancavelica and Potosí.[52]

52   Provisión del virrey del Perú don Melchor Portocarrero Lazo de la Vega, conde de la Monclova: Establece un nuevo régimen para la mita de Potosí, abandonando – por los numerosos inconvenientes que sobrevinieron en su aplicación – el régimen que había establecido su antecessor, don Melchor de Navarra y Rocafull, duque de la Palata. Lima, April 27, 1692. ABNB, Ruck 11, 23–37, 32; Lorenzo Mateo, indio del pueblo de San Juan de Challapata, ayllo Jilha, provincia de Paria, pidiendo se le exima de la mita de Potosí por estar enfermo y baldado. La Plata, February 23, 1736. ABNB, ALP Minas 126/16, 1; Registro de entrega de los indios de mita de las provincias de Porco, Canas y Canches, Chucuito, Chayanta, Paria, Asángaro y Umasuyo, por el capitán general de este servicio, don Juan José de Orense, a los dueños de minas e ingenios a quienes corresponden. Potosí, June 24–November 1, 1736. ABNB, Ruck 11, 1, 3, 15, 22; Liñan y Cisneros, 304; Audiencia de Lima, "Relación que la Real Audiencia de Lima hace al excelentísimo Sr. Marqués de Castel-Dosrius", 294–295; "Contestación al discurso sobre la mita de Potosí escrito en La Plata 9 de marzo de 1793 contra el servicio de ella," in "Una polémica en torno a la mita de Potosí a fines del siglo XVIII," María del Carmen Cortés Salinas, Revista de Indias, Vol. XXX, Nos. 119–122 (January–December, 1970), 173; Capoche, 135, 141; Matienzo, 136; Arana, 12–13; Alberto Crespo Rojas, "El reclutamiento y los viajes en la 'mita' del cerro Potosí," in La minería hispana e iberoamericana, Vol. I, na., (Leon, Spain: Catedra de San Isidoro, 1970), 468–469, 472–473; Iden, "La 'mita' de Potosí," in Revista Histórica, Vol. II2, (Lima, 1955–56), 171; Iden, La Guerra entre vicuñas y vascongados, 41–42; Pedro Arana, Las minas de azogue del Perú (Lima: Imprenta de "El Lucero," 1901), 12–13; Jeffrey Cole, The Potosí Mita, 1573–1700. Compulsory Indian Labor in the Andes (Stanford: Stanford University Press, 1985), 12, 17, 67; Brian M. Evans, "Census Enumeration in Late Seventeenth-Century Alto Perú: The Numeración General of 1683–1684," in Studies in Spanish Population History, D.J. Robinson, Ed. (Boulder: Westview Publishers, 1981), 27; Zimmerman, 183; Patiño Paúl Ortíz, 208, 210; Bakewell, Miners of the Red Mountain, 69, 71–73, 94–95; Donald L Wiedner, "Forced Labor in Colonial Peru," in The Americas, Vol.16, No. 4, (April, 1960), 364–367, 369; Lohmann Villena, Las minas de Huancavelica, 76, 104, 109, 139; Bargallo, La minería y metalurgía, 257; Jorge Basadre, "El Régimen de la Mita," in Letras, Vol. III (Lima: 1937), 325–328, 334–335, 339, 354; Patiño Paúl Ortíz, 208; Brown, "Worker's Health," 483. The Huancavelica mita conscripted Indians from the districts of Tarma, Jauja, Yauyos, Angaraes, Huanta, Castrovirreina, Lucanas, Vilcashuamán, Andahuaylas, Chumbivilcas, Cotabambas and Aymares (Kendall Brown, "Workers' Health and Colonial Mercury Mining at Huancavelica, Peru," in The Americas, Vol. V7. No. 4 (April, 2001), 470 and Lohmann Villena, Las minas de Huancavelica, 268–269, 355); Basadre, El Conde de Lemos, 122; Robins, Mercury, Mining and Empire, 33–34. For the Potosí mita, Indians were dispatched from the districts of Azángaro y Asillo, Cabana y Cabanilla, Canas y Canchis, Carangas, Chayanta, Chucuito, Cochabamba, Omasuyos, Pacajes, Paria, Paucarcolla, Porco, Quispicanches, Sicasica, Tarija, and Tinta (Evans, "Census Enumeration," 27–28). While Potosí and Huancavelica were the primary destination for mitayos assigned to mining, at times other mines received a mitayo allotment, including Castrovirreina, Caylloma, Carabaya, Pasco, Nuevo Potosí and Laicota.

A Ruinous Revelation: Huancavelica's Cinnabar

Having tackled the issues of labor, the miner's legal framework, and provision of hydraulic power in Potosí, and mill location, the final part of Toledo's equation centered on ensuring a reliable supply of mercury for silver refiners throughout the viceroyalty. This was relatively easily accomplished as Huancavelica had been producing mercury since 1564, primarily exporting their product to New Spain where the amalgamation system had already been adopted.

There are differing tales of how the Spanish learned of the secrets of Santa Bárbara hill. None are fully substantiated, but all revolve around Amador de Cabrera, a Spanish *encomendero*, or local official who was granted Indian labor by the Crown in exchange for Christianizing his charges and defending the realm. According to one account, Cabrera had attended a Corpus Christi celebration, during which he handed his hat for safe keeping to the child of an Indian named Gonzalo Navincopa, in the town of Chacas, present-day Acoria. Navincopa was either a *yanacona*, or servant, of Cabrera, or curaca of the town. In either case, the boy misplaced the hat, and, seeking pardon, Navincopa revealed the existence of the cinnabar laden Santa Bárbara mount. Others accounts suggest that a curaca named Juan Tumsuvilca, or Indians from the village of Conayca, had told Cabrera about the cinnabar deposit. Still another version asserts that an Indian informed Cabrera's wife of the deposits, prompting Cabrera to induce the Indian to divulge their location. What is known is that on November 1, 1563, Indians guided Cabrera to one of the primary cinnabar deposits. He subsequently named it Todos Santos, after the feast day upon which the revelation took place, although it would also become known as La Descubridora and Santa Bárbara. This simple event, high in the remote, windswept Andes, would transform how the region's indigenous population

---

(Diego García Sarmiento de Sotomayor, Conde de Salvatierra, García, "Relación del estado en que deja el gobierno de estos reinos el Conde de Salvatierra al Sr. Virrey Conde de Alba de Lliste," in *Colección de las memorias o relaciones que escribieron los virreys del Perú.* Vol. II. Ricardo Beltrán y Rózpide, Ed. (Madrid: Imprenta del Asilo de Huérfanos del S.C. de Jesús, 1921), 239, 253; Luis de Velasco, 110; Mendoza y Luna, "Relación del estado del Gobierno," 36; Patiño Paúl Ortíz, 210; Jorge Basadre, *El Conde de Lemos y su tiempo* (Lima: Editorial Huascaran, S.A., 1948), 132); Juan PerezTudela y Bueso, "El problema moral en el trabajo minero del indio (siglos XVI y XVII)," in *La minería hispana e iberoamericana.* Vol. I. N.A. Leon, Spain: Catedra de San Isidoro, 1970, 364, 366, 370; Brown, "Workers' Health," 483; Lohmann Villena, *Las minas de Huancavelica*, 104, 465; Cobb, *Potosí y Huancavelica*, 77; Audiencia de Lima, 294–295; Lohmann Villena, *Las minas de Huancavelica*, 140; Molina Martínez, *Antonio de Ulloa*, 109; Brown, "Worker's Health," 483; Vargas Ugarte, Vol. II, 200; Bakewell, Peter. Miners of the Red Mountain, 40–41.

interacted with their environment, and established an unprecedented, and toxic, link between Andean mining centers and the global economy. Cabrera soon traveled home to Guamanga and formalized his claim to the deposits on New Year's Day, 1564, excluding in the process whoever had revealed it to him.[53]

Not long after Cabrera was shown the deposit in the Santa Bárbara hill, an Indian named Fernando Huamán revealed the long-exploited site of Chacllatacana to Antonio Rodríguez Cabezudo, who also promptly registered it. Technically, the Crown owned all subsurface rights, having exercised this claim since 1128. In 1348, however, the Ordenamiento de Alcalá modified the arrangement by allowing individuals to own and exploit mineral deposits in exchange for payment of a tax. Cabrera had hardly registered his claim when, in February, 1564, an edict from Viceroy Diego López de Zúñiga reestablished the crown's ownership of the deposits and insisted on and the need for vice regal approval for all mining. This was the first challenge to Cabrera's and other's claims, however Viceroy López de Zúñiga died soon after. The audiencia, a court endowed with executive and limited legislative powers, ruled in his stead pending his replacement, and allowed cinnabar mining to continue.[54] Beyond their legal limbo, Huancavelica's miners had to contend with the fact that at this time there was no demand for quicksilver in the region. Faced with this, they began to export mercury to New Spain, where miners were increasingly employing the technique.

In the decade between the revelation of the cinnabar deposits to Cabrera and 1574, miners took advantage of the rich surface deposits through open pit mining. In these early days, however, mercury production did not always

---

53　Llano Zapata, 217; Cantos de Andrade, 303; Acosta, *Historia natural*, 243; Sebastián Lorente, *Historia del Perú bajo la dinastía Austriaca* (Paris: Imprenta de Poissy, ND), 292; N.A., "Memoria sobre la mina de azogue de Huancavelica," 85; na. "Memorial y relación de las minas de azogue del Pirú," 423; Caravantes, 134; Lizárraga, 134; Juan Solórzano y Pereyra, 314; Montesinos, Vol. II, 19; Antonio de Ulloa, *Viaje a la América meridional*. Vol. II, Andrés Samuell, Ed.(Madrid: Historia 16, 1990), 148; Arana, 6; Patiño Paul Ortíz, 22; Lohmann Villena, *Las minas de Huancavelica*, 21, 26; Salas Guevara, *Villa Rica de Oropesa*, 18–19; Lohmann Villena, *Las minas de Huancavelica*, 23; na, "Historia de la mina de Huancavelica," 66; Cobb, *Potosí and Huancavelica*, 33; Basto Girón, 223; Cook, *Demographic Collapse*, 203; Virgilio Roel Pineda, *Historia social y económica de la colonia* (Lima: Gráfica Labor, 1970), 101; Carrasco, 90–91; Contreras, *La ciudad del mercurio*, 19; Robins, *Mercury, Mining and Empire*, 22.

54　Daniel Alonso Rodríguez-Rivas, "La legislación minera hispano-colonial y la intrusión de labors," in *La minería hispana e iberoamericana. Ponencias del I coloquio internacional sobre historia de minería*. Vol. I. N.A. (Leon, Spain: Catedra de San Isidoro, 1970), 659; Patiño Paúl Ortíz, 23; Lohmann Villena, *Las minas de Huancavelica*, 31–34; Jakob, 273.

require excavation as pre-conquest miners had left abundant tailings from the Chacllatacana hill in the lavaderos which had been used to produce llimpi. This cinnabar was easily refined and had no cost of excavation. Moreover, beginning in 1574, mercury refined from lavaderos paid a royalty of twelve and one-half percent, instead of the normal twenty percent. To produce it, miners such as Cabrera had access to Indian labor by virtue of being an encomendero. Other miners had similar privileges, or benefitted from forced labor allotments, or contracted wage laborers.[55]

As mining and refining got underway, that which had been known as Sigsichaca, or the place of the grass bridge, transformed into a mining town. Like so many, it had a transient population and a small, often absentee, elite who eschewed Huancavelica's harsh and toxic environs. When not in Guamanga or Lima, they resided in the centrally located San Antonio and Santa Ana neighborhoods.[56] Despite the humbleness of the village, limited local market for its product, and uncertainties over their ownership of the mines, the miners had reason for optimism. Given Toledo's determination to promote the amalgamation system in Peru, they were confident that the demand for, and price of, their product would soon increase. Such hopes were reflected when the town was formally established by Francisco de Ángulo, who would serve as its first governor. In honor of Toledo's birthplace, the founders called it the Pueblo Rico de Oropesa on either August 4th or 5th, 1571. There was, however, some debate among the founders concerning the selection of the town's patron saint. Accentuating the intimate relationship between its inhabitants and the environment, on August 5 a heavy blanket of snow covered the town. This event quickly put an end to the discussion, and the founders settled on Our Lady of the Snows as the town's patron saint. Among its inhabitants at this time were about 400 Indian laborers, and about 300 Spaniards and Creoles, many of whom were part-time residents.[57]

---

55   Cobb, *Potosí y Huancavelica*, 35, 49; Bakewell, "Technological Change in Potosi," 78; Lohmann Villena, *Las minas de Huancavelica*, 29–30, 63, 109; Basadre, "El Régimen de la Mita," 329; Vargas Ugarte, Vol. II, 200; Wiedner, Donald L. "Forced Labor in Colonial Peru," 359; James Lockhart, *Spanish Peru, 1532–1560. A Social History.* (Madison: University of Wisconsin Press, 1994), 11; Lohmann Villena, *Las minas de Huancavelica*, 30; Robins, *Mercury, Mining and Empire*, 23.

56   Cantos de Andrade, 305; Carrasco, 104; Patiño Paúl Ortíz, 35–35, 44; Contreras, *La ciudad del mercurio*, 51; Robins, *Mercury, Mining and Empire*, 29.

57   Cantos de Andrade, 305; Llano Zapata, 217; Murua, 549; Caravantes, Vol. II, 69; Llano Zapata, 217; Zimmerman, 96; Patiño Paúl Ortíz, 24, 29; Carrasco, 98, 104; Patiño Paúl Ortíz, 24; Contreras, *La ciudad del* mercurio, 15; Lohmann Villena, *Las minas de Huancavelica*, 67;

The hopes of Huancavelica's miners soon, however, turned to despair and disbelief. Not only did Viceroy Toledo prohibit the export of mercury, but he ordered the expropriation of their mines. Henceforth, all mercury production, transport and sale would be a state monopoly. The bittersweet result for the miners was that they were transformed from mine owners into state subcontractors organized as a gremio. Beginning in March, 1573, the gremio started to work under the *asiento* system, in which they were to annually produce a fixed amount of mercury and receive a specified number of mitayos. After satisfying the quinto, they were to sell all quicksilver to the crown at a set price. This contract based system would endure, under varying terms, until 1779, when the crown began to subcontract all production to one person.[58]

The expropriation outraged the forty-three affected miners, and even prompted a failed assassination plot against Gabriel de Loarte, whom was tasked by the viceroy with enacting the measure.[59] Toledo reasoned, however, that not only could mercury be a source of revenue for the crown, but by controlling its allocation to silver miners, it would be possible to detect and limit contraband due to the relationship between the amount of mercury consumed and the amount of silver it yielded. By controlling mercury disbursement, royal officials could have an estimate of what a silver miner should bring in for taxation, operating on a rule of thumb that it required two kilograms of mercury to produce one kilogram of silver with average ore. When mercury was diverted to the black market, the Crown lost revenue from the quinto which was paid on mercury production, from the sale of the quicksilver, and from the tax which was not collected on the contraband silver produced as a result. Seeking to stifle illicit trade, crown contracts for the transport of mercury to Andean mining centers contained provisions regarding an acceptable loss of the metal during transport.[60]

Patiño Paúl Ortíz, 105; Robins, *Mercury, Mining and Empire*, 23, 29. Concerning the dispute over if the foundation occurred on August 4th or 5th, see Patiño Paúl Ortíz, 24–28.

58   Bakewell, "Technological Change in Potosí," 82–84; Zimmerman, 97–99; Cole, *The Potosí Mita*, 9; Lohmann Villena, *Las minas de Huancavelica*, 47, 72, 79, 88; Arthur Whitaker, *The Huancavelica Mercury Mine* (Boston: Harvard University Press, 1941), 10–11; Robins, *Mercury, Mining and Empire*, 29–30. For detailed information on the various asientos, see Caravantes, Vol. IV; Lohmann Villena, *Las minas de Huancavelica*, Zavalo, 225–228; and Patiño Paúl Ortíz, 111–128.

59   Cantos de Andrade, 304–305; Caravantes, Vol. IV, 135; Carrasco, 111–113; Bargallo, *La minería y metalurgía*, 256; Lohmann Villena, *Las minas de Huancavelica*, 46, 75, 78; Cobb, *Potosí y Huancavelica*, 41; Robins, *Mercury, Mining and Empire*, 29.

60   Marqués de Casa Concha, Relación de estado que ha tenido y tiene la real mina de Guancavelica. 1726, AGI, Lima 469, 47; Fernández de Cabrera, 90; Baltasar de la Cueva, Conde de

Among the unintended consequences of depriving the gremio of ownership of the mines was that they had no incentive to invest in their upkeep or
improvements. On the contrary, it simply reinforced practices which favored
meeting production quotas over sustainability, and resulted in the common
practice of mining pillars and other structural elements of in the mine. Such
practices were even more pronounced among second-tier miners who rented
from a gremio member. The result was an extraordinarily unhealthy and dangerous environment, exemplified by numerous deadly cave-ins.[61]

By 1573, Toledo had completed the framework for a regional transition to
amalgamation-based silver production, having addressed issues of labor, law,
and liquids. His reforms would have transformative effects, not just on silver
production, but on labor, Indian health, the environment, and the local and
global economies. The centrality of mercury to this process literally put Huancavelica on the map, and in Potosí, silver production quickly rebounded, aided
by the tons of tailings of lesser quality but now processable ores that had ac-
cumulated over the years.[62]

Although the Almadén mercury mine in Spain would sustain silver production in New Spain, and occasionally supplement supplies in the Andes, it could
not provision all of Spanish America. Quicksilver from Huancavelica was consequently vital for Andean silver production, and was on occasion remitted to
New Spain and Central America when production or transport from Almadén

---

Castellar, "Relacion general que el Excmo. Señor Conde de Castellar, Marqués de Malagon, Gentilhombre de la Cámara de su Majestad, de su Consejo, Cámara y Junta de Guerra
de Indias, Virey, Gobernador y Capitan General que fué de estos Reinos, hace del tiempo
que los gobernó, estado en que los dejó, y lo obrado en las materias principales con toda
distinction," in *Memorias de los virreyes que han gobernado el Perú durante el tiempo del
coloniaje español*. Vol. I. M.A. Fuentes, Ed. (Lima: Librería Central de Felipe Bailly, 1859),
186; Lizárraga, *Descripción del Perú*, 183; Cole, *The Potosí Mita*, 8; Molina Martínez, *Antonio
de Ulloa*, 108; Lohmann Villena, *Las minas de Huancavelica*, 62; Robins, *Mercury, Mining
and Empire*, 30, 70.

61   Memorial del capitán Don Pedro Gutíerrez Calderón, 1; Fernández de Castro, "El Conde
de Lemos da cuenta a S.M. del estado en que hallo el reino," 271; Caravantes, Vol. IV, 166;
Haenke, 265; Crosnier, 47–48; Fernández Alonso, 347; Patiño Paúl Ortíz, 89, 91–92, 130;
Carrasco, 243, 247; Lohmann Villena, *Las minas de Huancavelica*, 88, 90, 428, 451; Scribner,
5; Strauss, 561; Robins, *Mercury, Mining and Empire*, 56, 60. On the drawn-out efforts of
Pedro Subiela to map the mine and design repairs to it, see Kendall Brown, "El ingeniero
Pedro Subiela y el desarrollo tecnológico en las minas de Huancavelica (1786–1821)," in
Histórica. Vol. IIIo, No.1 (July, 2006): 165–184.

62   Francisco de Toledo, "Ordenanzas," 339–340; Lizárraga, 281; Bakewell, "Technological
Change in Potosi," 82–4, 93; Cole, *The Potosí Mita*, 18; Robins, *Mercury, Mining and Empire*,
28, 38–42.

faltered. For example, between 1592 and 1593, officials in Huancavelica dispatched about 253 metric tons of Huancavelica's mercury to Mexico. Following a cave-in in Almadén in 1750, between 1752 and 1755, around 1,148 metric tons were sent up the coast to Mexico, and about 207 metric tons to Guatemala. Following another cave-in in Almadén in 1758, Huancavelica sent 184 tons to Mexico between 1758 and 1759.[63]

"Ministers of Hell":[64] Governance, the Gremio and Greed

The environment and people's health in the region of Huancavelica were affected by the manner in which the city and the mines were administered. Because Huancavelica was so important to the empire, the Crown concentrated authority there to a greater degree than many other places. Reflecting this, the town's corregidor had exceptional powers; not only was he responsible for the civil affairs of his jurisdiction, but he also served as superintendent of the mine, placing him in charge of all issues concerning mining and refining there. His fiscal and political power were further enhanced with the relocation of the treasury offices from Guamanga to Huancavelica in 1578. As a result, he had control of the system through which the government purchased mercury from the gremio, as well as from their creditors who often received quicksilver in payment of debts. Huancavelica's ascendance caused friction with Guamanga, under whose jurisdiction it initially lay. In 1581, Huancavelica obtained municipal status, briefly freeing it from Guamanga's domain until the measure was reversed in 1586. In 1601, the municipality of Huancavelica was restored for the

---

63   Manso de Velasco, 165–172, 311–312; Arana, 81; Lohmann Villena, *Las minas de* Huancavelica, 334; Cobb, *Potosí and Huancavelica*, 49; Whitaker, 52, 54; Guillermo Mira Delli-Zotti, "El Real Banco de San Carlos de Potosí y la minería Altoperuana colonial, 1779–1825," in *La savia del imperio: tres estudios de economia colonia*, Julio Sánchez Gómez, Guillermo Mira Delli-Zotti and Rafael Dobado, Eds. (Salamanca, Ediciones Universidad Salamanca, 1997), 332; Rafael Dobado González, "Las minas de Almadén, el monopolio del azogue y la producción de plata en Nueva España en el siglo XVIII," in *La savia del imperio: tres estudios de economía colonial,* Julio Sánchez Gómez, Guillermo Mira Delli-Zotti and Rafael Dobado, Eds. (Salamanca, Ediciones Universidad Salamanca, 1997), 470; Matilla Tascón, Vol. II, 391–392; María Dolores Fuentes Bajo, "El azogue en las postrimerías del Perú colonial," in *Revista de Indias.* Vol. XLVI (January-June, 1986), 102; Pearce, "Huancavelica 1700–1759," 698; Molina Martínez, *Antonio de* Ulloa, 96; Lohmann Villena, *Las minas de* Huancavelica, 165; Arana, 81; Pearce, "Huancavelica 1700–1759," 698; Robins, *Mercury, Mining and Empire,* 10.

64   Memorial del capitán Don Pedro Gutiérrez Calderón, 1.

remainder of the colonial period, and in 1784 it would become the capital of the intendancy of the same name.[65]

Beginning in 1607, Huancevelica's corregidors, who usually served a two or three year term, were drawn from the *oidores*, or judges, on the Lima audiencia. They usually had little to no mining knowledge, but like the miners they were notoriously venal and had a very short term orientation. Recognizing this, and the ripple effect of contraband quicksilver on the royal coffers, in 1735 King Philip V (1683–1746) began to directly appoint governors there.[66]

Although the corregidor had jurisdiction over mining affairs, the day to day adjudication of mining issues was handled by an *alcalde mayor de minas*, or mining magistrate, to whom reported the *veedores*, or inspectors who had the responsibility of ensuring the safe and orderly operation of the mine. The veedores, who also served as the *alcaldes*, or mayors, of the town, were assisted in the mine by three *sobrestantes*, or assistant inspectors. These were responsible for the upkeep of stairs and scaffolds, general repairs, and ensuring that miners had candles and other necessary items. Under them were the *mayordomos*, or overseers, who kept vigil over a squad of Indian foremen who supervised laborers working in their zone. Although the veedores were to enforce mining regulations, their salary was largely paid by the gremio, thus creating impediments to enforcement. Administering the day to day affairs of the municipality was the *cabildo*, or city council, composed of the two *alcaldes/ veedores* and four regidores. Other officials included the *alguacil*, or sheriff, the *protector de los naturales*, or legal advocate for the Indians, and numerous scribes.[67]

---

65   Caravantes, Vol. II, 69; Carrasco, 115–116; Contreras, *La ciudad del mercurio*, 25, 28; Lohmann Villena, *Las minas de Huancavelica*, 117, 163; Brown, "La crisis financiera," 366; Robins, *Mercury, Mining and Empire*, 31; Carrasco, 28, 142; Patiño Paúl Ortíz, 33; Contreras, *La ciudad del mercurio*, 27–28. For a discussion of the Intendant system in Huancavelica, see Isabel M. Povea Moreno, *Retrato de una decadencia: régimen laboral y sistema de explotación en Huancavelica, 1784–1814*. Diss. (Granada: University of Granada, 2012), 52–61.

66   Ulloa, *Viaje*, Vol. II, 148; Navarra y Rocaful, 164–165; Bargallo, *La minería y metalurgía*, 261; Contreras, *La ciudad del mercurio*, 31–33, 38; Patiño Paúl Ortíz, 43; Molina Martínez, *Antonio de Ulloa*, 35; Lohmann Villena, *Las minas de Huancavelica*, 70; Whitaker, 22–25; Pearce, "Huancavelica 1700–1759," 682–683; Fernández Alonso, 370; Contreras, *La ciudad del mercurio*, 32; Carrasco, 115; Robins, *Mercury, Mining and Empire*, 31. For more on the town's civil administration, see Contreras, *La ciudad del mercurio*, 33–38.

67   Memorial del capitán Don Pedro Gutíerrez Calderón, 1; Mendoza y Luna, "Relación del estado del Gobierno," 41; Caravantes, Vol. IV, 146, 152; Cobb, *Potosí y Huancavelica*, 65; Lohmann Villena, *Las minas de Huancavelica*, 175–176, 448–449; Patiño Paúl Ortíz, 157–158; Brown, "Worker's Health," 471; Molina Martínez, *Antonio de Ulloa*, 67–68; Bakewell, *Miners*

It was not a yearning for royal service that motivated people to serve as civil officials in Huancavelica, instead, it was the prospect for graft. This was the only bright spot in an otherwise bleak scenario, as the town was isolated, small, toxic, and prone to discord. Illicit activities commonly involved participation in the numerous illegal mining schemes and trafficking in contraband mercury, both of which veedores and corregidores were well-placed to operate. The proximity of numerous silver mines in the region, such as Cerro de Pasco, Angarares and Lucanas, offered a ready market for quicksilver. For example, veedores routinely complemented their salaries by directing mitayos or others working under them to extract cinnabar from the various supports in the mines, or areas that had been closed for safety reasons. This cinnabar was then sold to refiners, yielding a profit for the official and the refiner while compounding the risk of cave-ins. Ironically, veedores who were to see to the integrity of the mine were damaging it, and mitayos who were drafted to work on shoring up the supports were instead mining them.[68]

Another time-tested means of revenue enhancement exploited the lack of coin in the town. Royal remittances were routinely delayed, with the result that both the gremio, and mitayos, would often have to wait to be paid for their mercury and labor. Even once money had arrived, the governor would often not release it, instead using his own funds (or those in the treasury) to purchase mercury from the cash-starved gremio at a steep discount before reselling it to the royal treasury at the price indicated in the asiento. Within the treasury offices, where mercury was purchased from the gremio and their creditors, a lack of oversight contributed to corrupt practices. In 1626, when King Phillip IV (1605–1665) ordered an audit of the treasury books, it was the first such investigation in thirty-three years.[69] By 1726, not much had changed

---

of the Red Mountain, 169–170; Whitaker, The Huancavelica Mercury Mine, 48; Lohmann Villena, Las minas de Huancavelica, 449; Haenke, 263; Patiño Paúl Ortíz, 66; Lohmann Villena, Las minas de Huancavelica, 449; Robins, Mercury, Mining and Empire, 37.

68   Memorial del capitán Don Pedro Gutíerrez Calderón, 1; Patiño Paúl Ortíz, 34, 43, 56, Contreras, La ciudad del mercurio, 46–47, 52–53, 57; Lohmann Villena, Las minas de Huancavelica, 116; Whitaker, 22; Crosnier, 56–57; Cueva, 186; Fernández de Cabrera, 90; N.A., Memoria sobre la mina de azogue, 117; Manso de Velasco, 159, 161; Sarmiento de Sotomayor, 245; Armendaris, 162; Navarra y Rocaful, 175; Molina Martínez, Antonio de Ulloa, 109; Kendall Brown, "La crisis financiera," 364–365, 375; Patiño Paúl Ortíz, 114; Lohmann Villena, Las minas de Huancavelica, 174, 254, 276, 284, 286; Bakewell, "Registered Silver Production," 84; "Fernández Alonso," 347, 362; Molina Martínez, Antonio de Ulloa, 109; 375; Robins, Mercury, Mining and Empire, 32, 110–112.

69   Carta de Pedro Camargo, 2; Marqués de Casa Concha, Relación del estado 53–56; Audiencia de Lima. "Relacion que la Real Audiencia de Lima hace," 295; Armendaris, 162, 362;

and Governor José de Santiago-Concha y Salvatierra, Marqués de Casa Concha referred to the "grave disorder" of Huancavelica's royal administration.[70]

Huancavelica's web of graft not only involved inspectors, corregidors, and treasury officials, but even viceroys, who, until 1735, appointed the corregidor. Despite the opportunities for graft and contraband, such activities were not without risk. Those convicted of trafficking mercury were subject to perpetual exile, the seizure of the quicksilver and one-half of their property, and the forfeiture of any mita allotment. On occasion, people were charged, but given the permeating climate of corruption, such cases tended to reflect political or personal rivalries.[71]

Huancavelica's Reach

Huancavelica's environmental history is intertwined with that of silver mining towns throughout the Andes, as much of the mercury volatized during the silver refining process deposited in the soils and waterways of Potosí, Oruro, Castrovirreyna and countless other places. Huancavelica's environmental impact was not just limited to the Andes, as its mercury was on occasion remitted to New Spain and Guatemala. Moreover, gaseous mercury can remain suspended in the earth's atmosphere, and travel thousands of kilometers, before settling in the oceans or landmass months later.[72]

Beyond the regional and global environmental effects were economic consequences. As with many mining towns, the city's high altitude limited what could be grown or raised there, spawning an expansive economic orbit. The combination of the isolation of the town and its thirst for goods made it an expensive place to live. Keeping the plaza's market supplied entailed transporting

---

Lohmann Villena, *Las minas de Huancavelica*, 285, 451–452; Brown, "La crisis financiera," 367; Patiño Paúl Ortíz, 138; Fernández Alonso, 358; Pearce, "Huancavelica 1700–1759," 676; Robins, *Mercury, Mining and Empire*, 31.

70   Marqués de Casa Concha, Relación del estado, 53.

71   Marqués de Casa Concha, Relación del estado, 47; Whitaker, 41, 43; Patiño Paúl Ortíz, 161; María C.N. Abrines, "El gobierno de Carlos de Beranger en Huancavelica (1764–1767)," in *Jahrbuch Für Geschichte Von Staat, Wirtschaft Und Gesellschaft Lateinamerikas*. Vol. III4 (1997), 107; Fernández Alonso, 347; Pearce, "Huancavelica 1700–1759", 677; Montesinos, Vol. II, 212; Fernández Alonso, 359–360, Molina Martínez, *Antonio de Ulloa*, 112; Robins, *Mercury, Mining and Empire*, 112.

72   USEPA, *Mercury Study*, 2–1, 2–4, 3–13.

almost everything in by pack animals such as mules and llamas, and by oxcart. The surrounding region provided extensive, if potentially contaminated, fodder for cattle, camelids, sheep, pigs and chickens, which provided the town with meat, milk, cheese, butter, lard and eggs. Farmers also cultivated tubers and quinoa, while fishermen brought fish from the region's high altitude lagoons.[73]

While most finished luxury products came via the vice regal capital of Lima, other prized products such as olives, rice, nuts and chocolate were produced regionally. From the more temperate climates of nearby Tayacaja, Lircay, Angaraes, Acobamba, and the Mantaro Valley, as well as in more distant Abancay and Andahuaylas, came vegetables, fruits and sugarcane, as well as the jams and candies sold on the plaza's market. The region of Castrovirreyna was a source of tubers, corn, cattle and goats, for both the mining community there as well as the region. From the warmer valleys of Huanta came the indispensable coca for the native residents. Under the Inca, the consumption of coca was restricted to the political elite, however its use became widespread after the conquest. When the leaves are chewed with bicarbonate powder, the release of alkaloids aid the absorption of carbohydrates in the body, suppress hunger and increase stamina.[74] In addition, coca serves as a religious and social currency,

73   Cantos de Andrade, 305, 308; Haenke, 267; Bueno, 72; *Descripción del virreinato del Perú*, 83; Contreras, *La ciudad del mercurio*, 79, 82–86, 88–89, 93, 95–96, 100; Lohmann Villena, *Las minas de Huancavelica*, 7, 69, 224; Patiño Paúl Ortíz, 45; Molina Martínez, *Antonio de Ulloa*, 143; Povea Moreno, *Retrato de una decadencia*, 285–287, 291, 296–297; Robins, *Mercury, Mining and Empire*, 32.

74   Thierry Saignes, "Capoche, Potosí y la coca: El consumo popular de estimulantes en el siglo XVII," in *Revista de Indias*, Vol. IV8, Nos. 182–183 (1988), 207, 213–214; Alfred Crosby, *The Columbian Exchange: Biological and Cultural Consequences of 1492* (Westport CT: Greenwood Press, 1972), 65–66, 71; Arzáns, Vol. II, 2, 268; Bartolomé Martínez y Vela, *Anales de la Villa Imperial de Potosí* (La Paz: Imprenta Artistica, 1939), 378; na. *Descripción del virreinato del Perú: Crónica inedita de comienzos del siglo XVII*. Boleslao Lewin, Ed. (Rosario, Argentina: Universidad Nacional del Litoral, 1958), 98–99; Lizárraga, 189, 191; Murua, 571; Caravantes,Vol. I, 146; N.A., "Descripción de la villa y minas de Potosí," 380; Cañete y Domínguez, 43; Toledo y Leiva, 176; Fernando de Torres de Portugal, Conde de Villardompardo, "Memoria gubernativa del Conde del Villardompardo." Lima, May 25, 1592 or 1593, in *Los virreyes españoles en America durante el gobierno de la casa de Austria. Perú*. Vol. I. Lewis Hanke and Celso Rodríguez, Eds. (Madrid: IMNASA, 1978), 209; Salinas y Córdoba, 264–265; Brading and Cross, 546; Luis Vilma Milletich Acosta and Enrique Tandeter, "El comercio de efectos de la tierra en Potosí. 1780–1810," in *Minería colonial Latinoamericana*, Dolores Avila, Inés Herrera and Rina Ortíz, Eds. (Mexico City: Instituto Nacional de Antropología e Historia, 1992), 137, 148; Cobb, "Supply and Transportation," 30; Zulawski, *They*

utilized in rituals which also promote group unity. Similarly, corn for *chicha*, a fermented alcoholic beverage widely consumed by the indigenous population, was cultivated in Jauja. Jauja also produced wheat, favored by the Spanish, who also obtained wines and brandies from vineyards and distilleries in the coastal regions of Ica and Pisco. In Guamanga, weavers and bakers produced textiles and wheat bread for Huancavelica's market, while debt peons labored on the region's haciendas to produce cereals and other products.[75]

Over the seventeenth century, the vigorous commerce in Huancavelica enabled the town's merchants to consolidate their economic position, ultimately surpassing that of the miners. Despite royal prohibitions to the contrary, mercury was often used as a currency as a result of the town's chronic lack of specie. It was this combination of ready money among the merchants, and need for capital among the miners, that allowed the former to purchase mercury at a discount. When funds ultimately arrived from Lima, the merchants would sell their mercury to the government at the official price, realizing a handsome profit in the process.[76]

The period from the 1520s, when European diseases first arrived in the Andes, to around 1650 were nothing less than a pachacuti for the indigenous people of the region of Huancavelica and elsewhere. Waves of influenza, typhus, smallpox, measles, plague and diphtheria ravaged the region, cutting the population in the region of Huancavelica by about sixty percent. The spread of such diseases was aided by the forced resettlement of Indians into reducciones,

---

Eat from Their Labor, 50–52; C. Sempat Assadourian, *El sistema de la economia colonial* (Mexico City: Editorial Nueva Imagen, 1983), 15, 51, 55, 63; Sempat Assadourian, Heraclio Bonilla, Antonio Mitre and Tristan Platt. *Minería y espacio económico en los Andes. Siglos XVI–XX* (Lima: Instituto de Estudios Peruanos, 1980), 13, 23–24; Crespo Rojas, *La Guerra entre vicuñas y vascongados*, 18; Saignes, "Capoche, Potosí y la coca," 208, 213–214, 227; Stein, 33; Robins, *Mercury, Mining and Empire*, 40, 42.

75   Juan de Matienzo, *Gobierno de Perú*, Guillermo Lohmann Villena, Ed.(Lima: Institut Fracncais D'Études Andines, 1967), 163–164; Bueno, 72; Cantos de Andrade, 305, 308; Haenke, 267; *Descripción del virreinato del Perú*, 83; Contreras, *La ciudad del mercurio*, 79, 82–86, 88–89, 93, 95–96, 100; Lohmann Villena, *Las minas de Huancavelica*, 7, 69, 224; Patiño Paúl Ortíz, 45; Molina Martínez, *Antonio de Ulloa*, 143; Povea Moreno, *Retrato de una decadencia*, 285–287, 291–292, 296–297; Henri Favre, "Evolución y situación de las haciendas," 237; Robins, *Mercury, Mining and Empire*, 32.

76   Carta de Pedro Camargo, 2; Memorial del capitán Don Pedro Gutíerrez Calderón, 2; Contreras, *La ciudad del mercurio*, 105–106, 108–109; Patiño Paúl Ortíz, 56; Pearce, "Huancavelica 1700–1759," 676; Robins, *Mercury, Mining and Empire*, 31.

where they could be monitored, controlled and catechized by their new masters. The reducciones also stripped Indians of their timeless connection to the geography of their original homes, which was laden with cosmological import. Disease and forced resettlement also upended the vertical archipelago system of exchange that many allyus had thrived upon for centuries, further forcing Indians into a cash economy. Extirpation campaigns sought to root out all indigenous forms of worship, physically destroying religious articles, flogging and shaming shamans, and castigating the natives. Despite this frontal assault, the resilience and tenacity of indigenous people, and their ancestral beliefs and practices were remarkable.

Nevertheless, those who survived did so in what Friar Salinas y Córdoba described as a "very unhappy and confused world."[77] They were subject to the often capricious authority of their curacas, who were often "interim" mestizo leaders who had no connection to or concern for their subjects. In addition to meeting the curaca's plethora of demands for labor and service, Andeans in the region of Huancavelica were also subject to the demands of their corregidor and priest. Beyond labor demands, corregidors coerced their charges to purchase unwanted goods through the reparto system, while clergy demanded extortionate payments for weddings and funerals, and derived additional revenues from saint's day celebrations and other commerce. All of these served as push factors for out-migration for both originarios and *forasteros*, or those who did not live in their birth-towns. Many migrated to Cuzco to seek wage work, while others sought employment in haciendas.[78] While the latter offered some protection from curacas and corregidors, it was at the expense of what liberty remained of the Indian, and his descendants who were bound to the estates for debt.

The revelation of Huancavelica's cinnabar deposits to Amador Cabrera in 1563, followed by the arrival of Viceroy Toledo in 1569 and the introduction of mercury amalgamation in silver production would further upend the indigenous world in the region of Huancavelica like no other in the Americas. Beyond the risks inherent in cinnabar mining and refining, the combination of highly concentrated civil power and entrenched networks of corruption that stretched to Lima exacerbated the risks to workers. Inspectors, overseers and others would routinely detail mitayos and wage workers to mine structural supports, resulting in cave-ins and death. Prior to the arrival of the Spanish,

---

77   Salinas y Córdoba, 287.
78   Cieza de Leon, 303; Ann Wightman, *Indigenous Migration and Social Change: The Foresteros of Cuzco, 1570–1720* (Durham: Duke University Press, 1990), 48–49, 149, 153.

native Andeans had limited excavation in the region to surface extraction. Following the conquest, they were conscripted to labor in the bowels of the Santa Bárbara Hill, or to load, operate and clean the town's many smelters. In this work, they interacted with their environment in new ways, ones that would literally permeate their bodies and was transmitted from mother to child. It is to the nature and scope of labor systems in Huancavelica that we now turn.

CHAPTER 3

"A Horrible Business": The Mita[1]

The Spanish did not introduce forced labor in Peru, the Inca and their pre-decessors had institutionalized it in the mita system. Like the Spanish, the Inca called up men for short-term service in infrastructure and public works projects. Under colonial rule, however, the mita expanded vastly in terms of the numbers of people pressed into service, the length of their bondage, and the scope of the tasks they were assigned. There was also an unbridgeable gap between how the mita was to operate in theory, and how it did in practice. Those subject to the colonial mita were originario tributary Indians between the ages of eighteen and fifty, with curacas, their heir, hilacata members, those who were serving the clergy, and the physically disabled exempt. In theory, no more than about one-seventh of a town's population was to be drafted in any given year.[2]

---

1 Carta del Virrey del Perú # 38, "Da cuenta a V. Md. de todo lo que en materia de azogue sea escrito." February 2, 1630, AGI, Indiferente general 1777, 3.

2 "Provisión del virrey del Perú don Melchor Portocarrero Lazo de la Vega, conde de la Mon-clova: Establece un nuevo régimen para la mita de Potosí, abandonando – por los numerosos inconvenientes que sobrevinieron en su aplicación – el régimen que había establecido su an-tecessor, don Melchor de Navarra y Rocafull, duque de la Palata." Lima, April 27, 1692. ABNB, Ruck 11, 23–37, 32; Lorenzo Mateo, indio del pueblo de San Juan de Challapata, ayllo Jilha, provincia de Paria, pidiendo se le exima de la mita de Potosí por estar enfermo y baldado. La Plata, February 23, 1736. ABNB, ALP Minas 126/16, 1; Registro de entrega de los indios de mita de las provincias de Porco, Canas y Canches, Chucuito, Chayanta, Paria, Asángaro y Uma-suyo, por el capitán general de este servicio, don Juan José de Orense, a los dueños de minas e ingenios a quienes corresponden. Potosí, June 24–November 1, 1736. ABNB, Ruck 11, 1, 3, 15, 22; Liñan y Cisneros, 304; Audiencia de Lima, "Relacion que la Real Audiencia de Lima hace al excelentísimso Sr. Marqués de Castel-Dosrius," 294–295; "Contestación al discurso sobre la mita," 173; Capoche, 135, 141; Matienzo, 136; Arana, 12–13; Crespo Rojas, "El reclutamiento y los viajes," 468–469, 472–473; Iden, "La 'mita' de Potosí," in *Revista Histórica*, Vol. II2, (Lima, 1955–56), 171; Iden, *La Guerra entre vicuñas y vascongados*, 41–42; Arana, 12–13; Cole, 12, 17, 67; Brian Evans, 27; Zimmerman, 183; Patiño Paúl Ortíz, 208, 210; Bakewell, *Miners of the Red Mountain*, 69, 71–73, 94–95; Wiedner, 364–367, 369; Lohmann Villena, *Las minas de Huancavelica*, 76, 104, 109, 139, 465; Bargallo, *La minería y metalurgía*, 257; Basadre, "El Régimen de la Mita," 325–328, 334–335, 339, 354; Iden, *El Conde de Lemos*, 122; Juan Perez Tudela y Bueso, "El problema moral en el trabajo minero del indio (siglos XVI y XVII)," in *La minería hispana e iberoameri-cana*. Vol. I. N.A. Leon, Spain: Catedra de San Isidoro, 1970, 364, 366, 370; Brown, "Workers' Health," 483; Cobb, *Potosí y Huancavelica*, 77; Robins, *Mercury, Mining and Empire*, 33–34.

© KONINKLIJKE BRILL NV, LEIDEN, 2017 | DOI 10.1163/9789004343795_005

While the mining mita is the most infamous form of service, mitayos were also assigned to public works projects such as constructing fortifications and churches, as well as for mail service and staffing *tambos*, or inns. Individuals who desired Indians as shepherds, for fieldwork and even private home construction in new towns could petition for mitayos as such work was broadly deemed in the public interest. The mita did not, however, include service in vineyards or on coca and tobacco plantations given the hot climate often associated with such products. Among the most feared form of mita was that destined for work in obrajes, especially those around Quito, where mitayos were often chained alongside convicts and child laborers. Routinely abused and deprived of their below-subsistence wage, many mitayos in obrajes became indebted, and permanently bound to the enterprise. Those who sought to regain their freedom through flight often found themselves in the complex's stockade instead.[3]

Technically, the mita was not slavery as the mitayos were not legally owned by the beneficiary, worked a limited term, and were to receive token compensation for both their time and travel. Instead, colonial authorities framed the draft as a tax on originario Indians for the privilege of utilizing community agricultural lands which had been recognized by the king. In contrast, as forasteros did not enjoy usufruct to community lands, they were exempt from

---

The Huancavelica mita conscripted Indians from the districts of Tarma, Jauja, Yauyos, Angaraes, Huanta, Castrovirreina, Lucanas, Vilcashuamán, Andahuaylas, Chumbivilcas, Cotabambas and Aymares (Kendall Brown, "Workers' Health and Colonial Mercury Mining at Huancavelica, Peru," in *The Americas*, Vol. V7. No. 4 (April, 2001), 470 and Lohmann Villena, *Las minas de Huancavelica*, 268–269, 355). For the Potosí mita, Indians were dispatched from the districts of Azángaro y Asillo, Cabana y Cabanilla, Canas y Canchis, Carangas, Chayanta, Chucuito, Cochabamba, Omasuyos, Pacajes, Paria, Paucarcolla, Porco, Quispicanches, Sicasica, Tarija, and Tinta (Evans, "Census Enumeration," 27–28). While Potosí and Huancavelica were the primary destination for mitayos assigned to mining, at times other mines received a mitayo allotment, including Castrovirreina, Caylloma, Carabaya, Pasco, Nuevo Potosí and Laicota. (Diego García Sarmiento de Sotomayor, Conde de Salvatierra, García, "Relación del estado en que deja el gobierno de estos reinos el Conde de Salvatierra al Sr. Virrey Conde de Alba de Lliste," in *Colección de las memorias o relaciones que escribieron los virreys del Perú.* Vol. II. Ricardo Beltrán y Rózpide, Ed. (Madrid: Imprenta del Asilo de Huérfanos del S.C. de Jesús, 1921), 239, 253; Luis de Velasco, 110; Mendoza y Luna, "Relación del estado del Gobierno," 36; Patiño Paúl Ortíz, 210; Jorge Basadre, *El Conde de Lemos y su tiempo* (Lima: Editorial Huascaran, S.A., 1948), 132).

3 Memorial del capitán Don Pedro Gutíerrez Calderón,3; Basadre, "El régimen de la mita," 334–336, 355; Iden, *El Conde de Lemos*, 122–123; Wiedner, 364, 366; Contreras, *La ciudad del mercurio*, 64; Patiño Paúl Ortíz, 133, 209; Robins, *Mercury, Mining and Empire*, 33. Ayans, 85; Robins, *Mercury, Mining and Empire*, 33, 167.

the mita. Not surprisingly, this distinction spurred massive internal migration, sometimes fleeing to areas outside of imperial control, to avoid the service. Beginning in 1631, the process was exacerbated as Indians were increasingly dispossessed of their lands as a result of the *composición de tierras*, or privatization of land through issuing of titles, further swelling the ranks of the forasteros.[4]

The rise of forasteros further debilitated communities which were already devastated by the demographic implosion. Just as when they were forced onto reducciones, those who became forasteros were shorn from ancestral lands to which they had a deep cosmological and personal connection, and lost their ability to participate in, and benefit from, the vertical archipelago system. For those left in their increasingly depopulated hometowns, the toxic and often mortal burden of the mining mita fell that much more heavily. Instead of having to serve every seven years, they increasingly found themselves pressed back into service in as little as two years, a tendency which only added to the cycle of out-migration.[5]

---

4   Sarmiento de Sotomayor, 254; Basadre, *El Conde de* Lemos, 140; Wightman, 151–152.

5   Representación de Dn. Mathias Chuquimanqui, Gobernador i Casique de Caquiviri, provincia de Pacajes, contra los españoles que en sus haciendas, hacen pasar indios sin pago de la tasa ni concurrencia a la mita. La Plata, September 28, 1743, ABNB, ALP Minas 149/14; Miguel Flores, Luis Flores y Santos García, indios del pueblo de Chayanta, provincia del mismo nombre sobre el exceso con que se trate de enviarlos a la mita de Potosí antes de los dos años de descanso. La Plata, September 1, 1757. ALP, Minas 127/12, 1; Simón Pérez, indio forastero en el pueblo de Asangaro, provincia del mismo number, sobre que se le asignen tierras de repartimiento en dicho puebo, como a mitayo de él. Asángaro, August 23, 1752. ABNB, ALP Minas 127/2, 1–2; Ayans, 38–39; Navarra y Rocaful, 239–240; Mesía Venegas, 120; Arias de Ugarte, "Carta a S.M. del nuevo oidor doctor Arias de Ugarte dando cuenta del estado en que halló la Audiencia de Charcas. Acompaña un memorial de los indios de su distrito en razón de los agravios que reciben. Pide que se tomen medidas para que no desembarquen extrajeros y gente perdida por el puerto de Buenos Ayres." Potosi, February 28, 1599. In *La audiencia de Charcas: correspondencia de presidentes y oidores. Documents del Archivo de Indias*. Vol. III, Roberto Levillier, Ed. (Madrid: Imprenta de Juan Pueyo, 1922), 364; Ayans, 35–88; Velasco, 109; Guaman Poma, 333; Pedro de Oñate, "Parecer del P. Pedro de Oñate sobre las Minas de Huancavelica 1629," in *Pareceres juridicos en asuntos de indias (1601–1718)*, Rubén Vargas Ugarte, Ed. (Lima: CIP, 1951), 143; Liñan y Cisneros, 305; Victorián de Villaba, "Vista del fiscal Victorián de Villaba, sobre los abusos de la mita," in *Vida y obra de Victorián de Villaba*, Ricardo Levene, Ed. (Buenos Aires: Instituto de Investigaciones Históricas, 1946), lvii; NA., "Pareceres de los Padres," 120–121; Caravantes, Vol. IV, 84; Fisher, *Silver Mines and Silver Miners*, 10; Paulino Castañeda Delgado, "Un capítulo de ética Indiana española: los trabajos forzados en las minas," in *Anuario de Estudios Americanos*, Vol. XXVII, (1970), 815; Wiedner, 368, 372; Roberto Choque Canqui, "El papel de los capitanes de indios de la provincia de Pacajes 'en el entero de la

The indigenous population in the region of Huancavelica was subject to various mita tasks and destinations, including periods when mitayos were remitted to Castrovirreyna, Caylloma, Carabaya, Cerro de Pasco, Vilcabamba, Laicota and the gold mines of Julcani and Zaruma. The mita assignment which inspired most dread, and caused the most flight, however, was that of toiling in Santa Bárbara's galleries and Huancavelica's smelters.[6] In order to provide the manpower necessary to produce enough mercury for the Andean silver mines, Viceroy Toledo initially designated twelve provinces surrounding Huancavelica to annually provide a rotating contingent of 3,289 men who would each serve for two months. Joining them were fifty *indios de plaza*, or plaza Indians, from the town of Capacmarca in Chumbivilcas. As they were not ordered to work in the mines or smelters, they served for six months in a variety of tasks working for civilian officials, priests, the mail service and the hospital. Indian residents in the town and immediate region were also drafted for a week's work of *faenas*, or diverse and generally unskilled tasks, such as upkeep of public works.[7]

---

mita' de Potosí," in *Revista Andina*, Vol. I, No. 1 (September, 1983), 120–121, 124; Teresa Cañedo-Arguelles Fábrega, *Potosí: La versión aymará de un mito europeo. La minería y sus efectos en las sociedades andinas del siglo XVII (La Provincia de Pacajes)* (Madrid: Editorial Catriel. 1993), 81, 83, 84–85; Iden, "Efectos de Potosí sobre la población indígena del Alto Perú. Pacajes a mediados del siglo XVII," in *Revista de Indias*, Vol. XLVIII, Nos. 182–183 (1988), 249; Nicolás Sánchez Albornoz, *Indios y tributos en el Alto Perú* (Lima: Instituto de Estudios Peruanos, 1978), 70; Iden, "Mita, migraciones y pueblos: variaciones en el espacio y en el tiempo Alto Perú, 5173–1692," in *Historia Boliviana*, Vol III, No. 1, (Cochabamba, 1983), 37–38; Buechler, *The Mining Society of Potosí*, 44; Saignes, "Las etnias de Charcas," 28; Cook, *Demographic Collapse,* 85, 210; Evans, "Census Enumeration," 27; Cole, *The Potosí Mita*, 26–27; Tandeter, *Coercion and Market*, 27; Fisher, *Silver Mines and Silver Miners*, 10; Robins, *Mercury, Mining and Empire*, 33, 35.

6   Memorial del capitán Don Pedro Gutíerrez Calderón, 2; Carta del Virrey del Perú # 38, 3; Expediente promovido por don Nicolás de Sarabia y Mollinedo asentista de la Real mina de azogue de Guancavelica sre que los corregidores y gobernadores de las provincias que tienen obligación de remitir mitas de indios para el trabajo de otra mina lo ejecuten sin dilación alguna y sre lo ocurrido con este motivo en la de Jauja a causa de haberse intentado cumplir con la remisión de mitayos aquel gobierno como se le mandó que avisa la resistencia que hicieron e insultos al cobrador enterador de mitas don Jacinto Maita. 1811. AGI, Indiferente 1335, 17–18; Sarmiento d Sotomayor, 239, 253–254; Mendoza y Luna, "Relación del estado del Gobierno," 36; Armendaris, 154; Fernández de Castro y Andrade, "Carta del Conde de Lemos," 155–156, 158; Patiño Paúl Ortíz, 196, 210–211; Basadre, "El Régimen de la Mita," 344; Iden, *El Conde de Lemos y su tiempo*, 132; Lohmann Villena, *Las minas de Huancavelica*, 225, 299, 382; Patiño Paúl Ortíz, 351; Robins, *Mercury, Mining and Empire*, 35–36.

7   "Provisión del virrey del Perú don Melchor Portocarrero Lazo de la Vega, conde de la Monclova: Establece un nuevo régimen para la mita de Potosí, abandonando – por los numerosos inconvenientes que sobrevinieron en su aplicación – el régimen que había establecido su antecessor, don Melchor de Navarra y Rocafull, duque de la Palata." Lima, April 27, 1692.

Inside the mine, there were generally three groups of workers. These included *barreteros*, who were *alquilas*, or wage laborers, and extracted ore by hammers and crowbars, and beginning in the 1730's, blasting. They were joined by mitayos known as *carguiches*, who ferried the ore to the surface, and *potabambas*, who were detailed to maintenance tasks. While carguiches were to work a shift of about nine hours, by the 1680s a quota-based system prevailed in which the worker was required to carry out at least forty *punchaos*, or loads, during their term of service, each of which were evaluated for size and quality.[8] One punchao was equal to a leather bag measuring about one and one half meters tall and one half meter wide, however miners often distributed bags larger than this. Although mitayos were expected to complete one punchao a day, this was not always possible depending how far they had to travel to gather and transport a load, and the quality of the ore. As a result, although mitayos were, in theory, to work forty days over two months, in order to meet their quotas many had to work during their supposed "down time" and or extend their term in Huancavelica. Once at the pithead, the ore was transported to the smelters on llamas, by workers known as *chacaneas*.[9]

ABNB, Ruck 11, 23–37, 32; Lorenzo Mateo, indio del pueblo de San Juan de Challapata, ayllo Jilha, provincia de Paria, pidiendo se le exima de la mita de Potosí por estar enfermo y baldado. La Plata, February 23, 1736. ABNB, ALP Minas 126/16, 1; Registro de entrega de los indios de mita de las provincias de Porco, Canas y Canches, Chucuito, Chayanta, Paria, Asángaro y Umasuyo, por el capitán general de este servicio, don Juan José de Orense, a los dueños de minas e ingenios a quienes corresponden. Potosí, June 24–November 1, 1736. ABNB, Ruck 11, 1, 3, 15, 22; Liñan y Cisneros, 304; Audiencia de Lima, "Relacion que la Real Audiencia de Lima hace al excelentísimso Sr. Marqués de Castel-Dosrius," 294–295; "Contestación al discurso sobre la mita de Potosí," 173; Capoche, 135, 141; Matienzo, 136; Arana, 12–13; Crespo Rojas, "El reclutamiento y los viajes," 468–469, 472–473; Iden, "La 'mita' de Potosí," in *Revista Histórica*, Vol. II2, (Lima, 1955–56), 171; Iden, *La Guerra entre vicuñas y vascongados*, 41–42; Arana, 12–13; Cole, 12, 17, 67; Brian Evans, 27; Zimmerman, 183; Patiño Paúl Ortíz, 208, 210; Bakewell, *Miners of the Red Mountain*, 69, 71–73, 94–95; Wiedner, 364–367, 369; Lohmann Villena, *Las minas de Huancavelica*, 76, 104, 109, 139; Bargallo, *La minería y metalurgía*, 257; Basadre, "El Régimen de la Mita," 208; Perez Tudela, 364, 366, 370; Brown, "Workers' Health," 483; Cobb, *Potosí y Huancavelica*, 77; Contreras, *La ciudad del mercurio*, 49, 64; Patiño Paúl Ortíz, 133, 209; Molina Martínez, *Antonio de Ulloa*, 82; Robins, *Mercury, Mining and Empire*, 52.

8 Marqués de Casa Concha, Relación del estado, 17; Caravantes, Vol. IV, 57; Navarra y Rocaful, 175; Bargallo, "La mineria y metalurgía," 262; Lohmann Villena, *Las minas de Huancavelica*, 311; Pearce, "Huancavelica 1700–1759," 691; Bargallo, *La minería y la metalurgía*, 262–263; Patiño Paúl Ortíz, 158; Molina Martínez, *Antonio de Ulloa*, 83; Cook, *Demographic Collapse*, 205; Zimmerman, 183; Robins, *Mercury, Mining and Empire*, 52.

9 Armendaris, 153; Navarra y Rocaful, 175; Torres de Portugal, 211–212; Lohmann Villena, *Las minas de Huancavelica*, 433; Basadre, *El Conde de Lemos*, 156; Robins, *Mercury, Mining and Empire*, 32.

When the mita system was instituted, most mitayos who worked in Huancavelica were, in theory, to be paid one and one quarter reales a day, in addition to receiving a kilogram of meat and two-thirds of a bushel of corn each month. They were also to receive a *porina*, or payment for the travel time to and from Huancavelica, however this was often not paid. Until 1632, one-half real a week was deducted from the mitayo's wages to offset the costs of the salaries of the protector de los naturales, interpreters and other mine officials.[10] Mine workers, including mitayos, were also granted the privilege of exchanging a *guasacho*, or chunk of ore, at the pithead, to vendors selling food or chicha, a practice which also attracted wage laborers. Draft laborers were to be paid in silver on Sundays in Huancavelica's main plaza in the presence of the corregidor, protector de los naturales and a clergyman, who were to inquire if they had received their allotment of food, were provided with candles, or had been mistreated.[11]

Despite its coercive nature and corrupt administration, Toledo's mining regulations did provide some, at least theoretical, protections for mitayos. These included provisions that mitayos not be made to work in climates different from those of their hometown, and that they begin work an hour and a half after sunup, have one hour for lunch, and work no later than sundown. During the winter months, those working the wet lavaderos were only to labor for six hours beginning at ten in the morning. The toxicity of refining mercury was

---

10   Mendoza y Luna, "Relación del estado del Gobierno," 41; Caravantes, Vol. IV, 139; Haenke, 138; Guaman Poma, 333; Martín José de Mugica, "Abusos de varias clases de mitas y carácter perezoso del Indio," in "Las mitas de Huamanga y Huancavelica," Luis Basto Girón, in *Perú Indígena*, Vol. V, No. 13 (December, 1954): 223; Patiño Paúl Ortíz, 157, 168; Bargallo, *La minería y la metalurgía*, 262–263; Lohmann Villena, *Las minas de Huancavelica*, 104, 155, 382; Basadre, *El Conde de Lemos*, 156; Iden, "El regimen de la mita," 349; Brown, "La crisis financiera," 364–365; Cobb, *Potosí and Huancavelica*, 76–77; Patiño Paúl Ortíz, 157; Bakewell, *Miners of the Red Mountain*, 170–171. In 1590, the wage was increased to two and one-half reales each day, and in 1650 it was raised to three and one half reales a day, less than one-third that of an alquila. There was, however, some differentiation based on specific tasks. For example, carguiches received one real for each load brought to the surface, while potabambas received four reales per day. For more information on mitayo wages, see See Caravantes, Vol. IV, 139; Cobb, *Potosí y Huancavelica* 65.

11   Memorial del capitán Don Pedro Gutíerrez Calderón, 3; Marqués de Casa Concha, Relación del estado, 10; Patiño Paúl Ortíz, 157; Bargallo, "La mineria y metalurgía," 263; Lohmann Villena, *Las minas de Huancavelica*, 439, 441; Basadre, *El Conde de Lemos*, 156; Tandeter, "Mineros de Week-end," 37, Iden, *Coercion and Market*, 111; Zulawski, "Wages, Ore Sharing and Peasant Agriculture," 412; Basadre, "El regimen de la mita," 349; Vargas Ugarte, 202; Bargallo, "La mineria y metalurgía," 263; Patiño Paúl Ortíz, 157; Lohmann Villena, *Las minas de Huancavelica*, 82; Wiedner, 369; Robins, *Mercury, Mining and Empire*, 61, 97.

well-recognized at the time, and was addressed through special regulations. These included stipulations that, prior to opening the smelters, they were allowed to cool for fourteen hours if ichu had been used as a fuel, or twenty-four hours if wood had been used. Technically, Indians were not to perform this task, but rather the owner of the refining operation, or his slave. Finally, mitayos injured on the job were to receive two reales a day while in the hospital, and allowed to convalesce for ninety days while those who suffered mercury poisoning were to be compensated with fifty pesos.[12]

These regulations served more to assuage the consciences of colonial officials and priests than to alleviate the travails of mitayos. As one writer explained, it was "a world where illegality... [had] the force of law."[13] The lack of efficacy of such protective regulations is hardly surprising in a context where the predominant view of the natives was that they were barely human and did not resent physical abuse. This, combined with the centrality of mercury to the imperial economy and pervasive corruption among civil and ecclesiastical officials, fostered a climate of physical abuse among those engaged in cinnabar mining and mercury production.[14]

---

12   Gaspar de Carvajal, Alonso de la Cerda, y Miguel Adrián al Rey, March 17, 1575, AGI, Lima, 314, 1; Caravanes, Vol. IV, 147; Toledo, "Ordenanzas," 340, 343–344; Zimmerman, 183; Lohmann Villena, *Las minas de Huancavelica*, 175–176; Cobb, *Potosí and Huancavelica*, 52, 71; Robins, *Mercury, Mining and Empire*, 37–38.

13   Ayans, 35, 60; Arias de Ugarte, 359, 366; Capoche, 160; Mendoza y Luna, "Relación del estado del Gobierno," 40; Testimonio de los informes que a instancia del doctor don Victorián de Villava, fiscal de esta Real Audiencia, expidieron don Francisco de Viedma, gobernador intendente de Puno; el doctor Felipe Antonio Martínez de Iriarte, cura propio de la doctrina de Chaqui, partido de Porco, y vicario pedáneo de Potosí; y el doctor don José de Osa y Palacios, cura propio que fue de la doctrina de Moscarí, partido de Chayanta, sobre los perjuicios que a los pueblos de indios de dicha circunscripción se siguen de la mita de Potosi. La Plata, November 24, 1794. ABNB, ALP. Minas 129.8, 3; Expediente instruído con las representaciónes de varios indios mitayos del pueblo de Capinota, provincia de Cochabamba, sobre los tributos que indebidamente se les exigen por los años que sirvieron la mita de Potosí, October 27, 1792. ABNB, ALP Minas 129/2, 2, 7; N.A., "Pareceres de los Padres de la Compañía de Jesús de Potosi," 119; Sarmiento de Sotomayor, 247; Torres de Portugal, 210; Cañete y Domíngez, 105, Navarra y Rocaful, 240; Balthasar Ramírez, 347; Mesía Venegas, 105; Armendaris, 153; Arana, 13; Mugica, 223; Lohmann Villena, *Las minas de Huancavelica*, 177; Jorge Basadre, "El Régimen de la Mita," in *Letras*, Vol. III (Lima: 1937): 349; Tandeter, *Coercion and Market*, 57–58; Bakewell, *Miners of the Red Mountain*, 90–91, 97, 104–105; Crespo Rojas, "La 'mita' de Potosí," 175, 178; Wiedner, 368; Cole, *The Potosí Mita*, 24–25; Cobb, *Potosí y Huancavelica*, 65; Saignes, "Las etnias de Charcas," 74. "un mundo donde la ilegalidad tiene fuerza de ley."

14   Matienzo, *Gobierno de Perú*, 17; Robins, *Mercury, Mining and Empire*, 47, 181.

Given the nature of their work, and the near uselessness of the regulations designed to protect them, many mitayos rightly viewed their service as a death sentence.[15] Fear turned to anguish when mita contingents were organized by the curacas every two months. Just prior to departure, the corregidor mustered the departing mitayos on the village plaza before dispatching them under the supervision of a *capitán enterdaor de mita*, or mita captain who had the fiscal responsibility to ensure the laborers arrived and completed their terms. The departure was a lugubrious affair, punctuated by melancholy tunes, tearful goodbyes and often the sound of chains as Indians were collared and shackled to prevent flight.[16]

For some, there was a middle way between service and flight, which consisted of essentially paying a ransom to escape being designated as a mitayo. If the prospective mitayo had the means, he could pay the curaca, who in turn would theoretically use the funds to compensate the beneficiary of the mita allotment in Huancavelica and cover the cost of a wage laborer in lieu of the mitayo. This practice was referred to as sending *indios en plata*, or Indians in silver. To accomplish this, many Indians would sell or pawn their livestock or other possessions.[17] According to Friar Salinas y Córdoba, who wrote around 1630, to avoid the service, men would even rent out their children or wives for between fifty and sixty pesos, and as a result "all those towns

---

15   Carta del Virrey del Perú # 38, 3; "Carta del Cura Vicario del Tomavi, provincia del Porco, a la Rl. Aud de la Plata." Tomavi, November 13, 1616. ABNB, CACh 728, 5; Montesinos, Vol. II, 205; Pedro Fernández de Castro y Andrade, Conde de Lemos, "Carta del Conde de Lemos a S.M. sobre la Mita de Potosí," in *Pareceres juridicos en asuntos de indias (1601–1718)*, Rubén Vargas Ugarte, ed. (Lima: CIP, 1951), 155; Salinas y Córdoba, 277–278; Cobb, *Potosí y Huancavelica*, 72; Robins, *Mercury, Mining and Empire*, 47.

16   Carta del Virrey del Perú # 38, 3; Sarmiento de Sotomayor, 246; NA., "Pareceres de los Padres de la Compañía de Jesús de Potosi," 118; Arzáns de Orsúa, Vol. III, 68; Fernández de Castro y Andrade, "Carta del Conde de Lemos," 155, 158–159; Salinas y Córdoba, 295–296; Brown, "Worker's Health," 474; Buechler, *The Mining Society of Potosi*, 137; Basadre, *El Conde de Lemos*, 133; Basadre, "El Régimen de la Mita," 344; Crespo Rojas, "El reclutamiento y los viajes," 474; Crespo Rojas, "La 'mita' de Potosí," 176, 179–180; Wiedner, 368; María del Carmen Cortés Salinas, "Una polémica en torno a la mita de Potosí a fines del siglo XVIII," in *Revista de Indias*. Vol. XXX, Nos. 119–122 (January-December, 1970), 178; Choque Canqui, 117, 123; Lohmann Villena, *Las minas de Huancavelica*, 276; Tandeter, *Coercion and Market*, 68–69; Robins, *Mercury, Mining and Empire*, 47–48.

17   Sarmiento de Sotomayor, 246; Fernández de Castro y Andrade, "Carta del Conde de Lemos," 165; Luis Valcárcel, "El 'Memorial' del Padre Salinas," in *Memorial de las Historias del nuevo mundo Piru*. Fray Buenaventura Salinas y Córdoba (Lima: Universidad Nacional Mayor de San Marcos, 1957) xi; Lohmann Villena, *Las minas de Huancavelica*, 356; Scribner, 12; Robins, *Mercury, Mining and Empire*, 47.

are filled with mestizo bastards...living witnesses to the rapes...and violence of so many soulless ones."[18]

Alternatively, the mitayo could deal directly with the miner to whom he was assigned, and pay a fee to be released from service. In this case, he was considered an *indio de faltriquera*, or an Indian in the pouch, because the recipient pocketed the commutation fee of between thirty and fifty pesos.[19] As in the case of the indio en plata, the funds were to be dedicated to hiring a free wage laborer in the stead of a mitayo. In practice, however, curacas commonly retained a part or all of the fee they collected and sent fewer men, while miners either used the funds to hire alquilas or simply kept the money if they did not

---

18   Salinas y Córdoba, 296; Carta del Virrey del Perú # 38, 3.

19   Don José de la Rúa, protector de naturales de Potosí, sobre el indebido cobro de 52 pesos anuales que el subdelegado del partido de Carangas, pretende llevar con el título abusivo de rezagos de mita, a cada indio de los que, estando repartidos para dicho servicio, no asistieron a él en virtud del indulto concedido últimamente por las cortes extraordinarias. Potosí, November 28, 1814. ABNB, ALP Minas 130.10, 1; Testimonio de los obrados relativos a los mitayos que deben mandar a Potosí, para el trabajo de minas del lugar de Sicasica. Potosí, 1759. CNMAH, CGI/M-65/17, 1; Capítulos puestos por los indios del pueblo de Calacoto, provincia de Pacajes, contra su cacique don Juan Machaca, y José Rivera, escribano publico de dicha provincia, sobre exacciones, malos tratamientos con motivo de la mita de Potosí, defraudaciones de tributes, etc. 1747–1750. ABNB, ALP Minas 126.20, 3; A los oficiales reales de Potosí: Se ha ordenado al virrey del Perú que haga correr la mita solamente en las 16 provincias primitivamente afectadas y ya no en las que después se agregaron con otras disposiciones sobre este servicio. Seville, October 22, 1732. ABNB, Cédulas Reales 545, 2; Provisión circular expedida por esta Real Audiencia pra que los gobernadores de inidos cesen en la costumbre de substituir a los contribuyentes acomodados con otros infelices en el servicio de la mita de Potosí. La Plata, 1773. ABNB, ALP Minas 127/22, 2; Avisos muy importantes y noticias muy particulares que de un bien intencionado y deseoso del mayor revicio... sobre el mayor regimen, estableciemiento y gobierno de la real mita. Potosí, January 20, 1762. ABNB, ALP Minas 151/10, 3; Testimonio de los informes, 6, 9; Capítulos puestos por Pedro Pirua y otros indios del pueblo de Chayanta, provincia del mismo nombre, contra su gobernador don Sebastián Auca, sobre defraudación de mitayos y otros excesos. Potosí, 1757–1758. ABNB, ALP Minas 127/14, 1–4; Miguel Flores, Luis Flores y Santos García, 1; Los indios de las parcialidades de Laymes, Cullpas, Chayantacas, Carachas, Puracas y Sicoyas, de la provincia de Chayanta, sobre extorciones a contribuyentes de la mita y otros excesos de los caciques. La Plata, 1797–1799. ABNB, ALP Min 129/16, 28; Haenke, 136; Fisher, *Silver Mines and Silver Miners*, 10; Bakewell, *Miners of the Red Mountain*, 124; Tandeter, "Forced and Free Labor," 123, 128; Iden, *Coercion and Market*, 67; Sánchez Albornoz, *Indios y tributos*, 103; Cole, 63; Crespo Rojas, "La 'mita' de Potosí," 176; Povea Moreno, "Entre la retórica y la disuación," 206; Robins, *Mercury, Mining and Empire*, 36.

have active mines or smelters. In the end, those who actually served the mita
were the ones who did not have money to avoid the service.[20]

Although mitayos were forbidden to live with their families during their mita
term, many traveled together so that the wife and children could help support
the mitayos during his term of service. The march to Huancavelica was not just
of people, for mitayos also brought livestock to carry, and ultimately serve as,
food during their stay.[21] Upon arrival, mitayo families were split up, with the
mitayos assigned, based on their hometowns, to large, dark, cold hovels, which
one chronicler described as fit for "rented slaves."[22] Most mitayos were ordered
to live in the vicinity of a concentration of smelters in what is now the San
Cristóbal neighborhood. While mitayos were designated as carguiches, potabambas, or smelter operators, their accompanying wives and children engaged
in petty commerce, served as domestic servants or assisted their husbands and
fathers in refining.[23] In 1623, one contemporary described the Indian workers
of Huancavelica as "poor, poorly clothed, their homes as if robbed and...their
wives and sons thin and ill."[24]

Despite the efforts of viceroys and corregidors, rarely did full contingents of
mitayos arrive. Flight, disease, and death all took their toll on the population,
and officials only reluctantly and belatedly adjusted levies so that they would
have some relation to demographic realities. The early levy of 3,289 men was
unsustainable, and in 1586 Viceroy Garcia Hurtado de Mendoza, Marqués de
Cañete, reduced it to 2,274, while Viceroy Luis de Velasco further pared it to

20   Provisión de don Melchor de Navarra y Rocaful, Duque de la Palata virrey del Perú: En
      cumplimiento de la Real Cédula de 1676.12.08, se extiende la obligación de dar indios
      mitayos para las minas e ingenios de Potosí a los pueblos que hasta ahora estaban ex-
      entos y se señala el orden general que ene ése y otros puntos ha de tener este servicio.
      Lima, December 2, 1688. ABNB Ruck 1/4: 451; Memorial que el licenciado Gaspar González
      Pavón escribió, de Potosí a España, a don Gómez Dávila, corregidor provisto de dicha Villa:
      Refiérle principalmente al régimen de la mita. Potosí, January 25, 1658. ABNB, ALP Minas
      125/14, 1–10; Los indios de las parcialidades de Laymes, Cullpas, Chayantacas, Carachas,
      Puracas y Sicoyas, 27; Sánchez Albornoz, *Indios y tributos*, 102; Robins, *Mercury, Mining
      and Empire*, 36.
21   Salinas y Córdoba, 264; Capoche, 135; Ramírez, 348; Joseph Baquijano, "Historia del descu-
      brimiento del cerro de Potosí, fundación de su Imperial Villa, sus progresos y actual es-
      tado," in *Mercurio Peruano*, 7 (1793), 37; Crespo Rojas, "La 'mita' de Potosí," 172–173; Cobb,
      Potosí y Huancavelica, 63; Strauss, 564; Robins, *Mercury, Mining and Empire*, 48.
22   Memorial del capitán Don Pedro Gutiérrez Calderón, 1–2. "esclavos alquilados."
23   NA, "Memoria sobre la mina de azogue," 129–130; Strauss, 564; Povea Moreno, *Retrato de
      una decadencia*, 232; ; Brown, "Worker's Health," 479–480; Robins, *Mercury, Mining and
      Empire*, 57–60, 72, 136, 181.
24   Memorial del capitán Don Pedro Gutíerrez Calderón, 3. "pobres mal vestidos, sus casas
      como robadas y últimamente ellos, sus mujeres e hijos flacos y enfermos."

1,600 in 1604. By 1621, flight and death had so riddled the provinces of Azángaro, Aymaraes, Vilcas, Guachos y Ananguncas and Soras y Lucanas that outgoing Viceroy Francisco de Borja acknowledged that these regions were unable to provide the number of mitayos demanded by the government. By 1630, approximately 1,000 of a further reduced levy of 1,400 mitayos arrived. Fifteen years later, despite calling up only 620 mitayos, less than half of that number mustered for service. The free fall continued, as evidenced by the arrival of 286 mitayos in 1684, and only forty-four in 1685.[25]

Given the effects of epidemics in the region, depopulation unquestionably limited the ability of communities to send the required number of mitayos. As one contemporary related in 1623, "the caciques cannot deliver the entire [levy]...because many of their Indian subjects have left and the rest... were killed in the mine." Those whom had fled found refuge in "remote and inexpungible" places beyond the reach of civil authority.[26] To this was added deaths as a result of mining and refining, which appear to be small in relation to the die off from epidemics. Nevertheless, as a physician at Huancavelica's hospital emphasized, "in the hollow of the mine are hidden many deaths."[27] While those who perished in cave-ins and other accidents often remained buried in the mines, other workers were able to make it home, where they died or were incapacitated as a result of mercury and arsenic poisoning and silicosis. Their sorry state made a strong impression on prospective mitayos, who were all the more inclined to flee. These different causes had the same outcome, the depopulation and desolation of the region, which was clear to contemporaries.

For example, in 1604 the protector of the Indians Damián de Jeria explained how

> the mortality of Indians has principally occurred in the valley of Jauja, caused by work in the shafts, and [this valley was]... among the best and most populated of Peru... and today... it is destroyed and almost razed and almost without people and the few that are there are so poor... such poverty has been caused by the illnesses and deaths of so many Indians, and ransoms to be free of the shafts.[28]

---

25 Montesinos, 94; Brown, "Worker's Health," 469; Borja, 85; Fisher, *Silver Mines and Silver Miners*, 10; Arana, 12; Lohmann Villena, *Las minas de Huancavelica*, 156, 189, 304, 309–310, 354, 378, 385, 427, 431, 460; Brown, "Worker's Health," 469, Robins, *Mercury, Mining and Empire*, 61–62.

26 Memorial del capitán Don Pedro Gutíerrez Calderón, 1–2.

27 Emetherio Ramírez de Arellano, 2.

28 Jeria, 3.

In a 1623 letter to Philip IV, a former official whom had spent time in Huancavelica asserted that "if this working of the mercury mines lasts thirty years, for seventy leagues around there will be no Indians left, and the cause of this will be because this mine is more dangerous than others of mercury."[29] Likewise, Viceroy Melchor Liñan y Cisneros noted the depopulation of Huancavelica's mita-subject provinces, noting that only 354 mitayos had arrived in 1681. In his view, the cause was none other than the Indian's "natural laziness" as opposed to fear and resentment of working in mercury production.[30] Eight years later, Viceroy Navarra y Rocaful advanced a more nuanced view, noting the importance of mita flight in depopulating the region. In the early seventeenth century, the chronicler Francisco López de Caravantes recognized that the mita did not just affect population as a result of flight and death, but through the separation of husbands and wives during the mita term.[31] In 1657, an oidor in Lima's audiencia traveled to the village of Santa Lucia, located in the mita-subject province of Lucanas. Writing to Philip IV, he explained how only women tended the fields, every man having fled save one who served as the sacristan, and hence was exempt from the mita.[32]

Three years later, Friar Buenaventura Salinas y Córdoba likewise noted the depopulation of the region, spurred by the mita. He averred that "Here they cry tears of blood and lament [in] the valleys of Jauja [and] the provinces of los Yauyos... because their Indians have ended up in forced and violent oppression, work and agony."[33] Like many contemporaries, Friar Reginaldo de Lizárraga attributed much of the depopulation to excessive alcohol consumption, as opposed to the mercury mita, while fearing for a "land without inhabitants, and the realm without vassals."[34]

Beyond death and relocation, some Indians essentially "disappeared" from the ranks of those subject to the mita. One way to achieve this was to become a debt peon on a hacienda, where an Indian was protected from both corregidor

---

29   Carta de Pedro Calderón, 1.

30   Liñan y Cisneros, 309–310; Lohmann Villena, *Las minas de Huancavelica*, 423.

31   Caravantes, Vol. IV, 214; Navarra y Rocaful, 239–240; Fisher, *Silver Mines and Silver Miners*, 10; Cañete y Domínguez, 115–116; Cook, *Demographic Collapse*, 251; Robins, *Mercury, Mining and Empire*, 35–36, 192. Regarding the deleterious effect of the mita on reproduction, see also Fernández de Castro, "Carta del Conde de Lemos," 158; Mesia Venegas, 100; Ayans, 43; NA. "Pareceres de los Padres de la Compañía de Jesús de Potosí," 118; Lohmann Villena, *Las minas de Huancavelica*, 299; Vargas Ugarte, Vol. II, 205–206.

32   Patiño Paúl Ortíz, 242; see also Oñate, 141–143.

33   Salinas y Córdoba, 278; Brown, "Worker's Health," 474.

34   Reginaldo de Lizárraga, *Descripción del Perú, Tucumán, Río de la Plata y Chile* (Buenos Aires: Union Académique Internationale/ Academia Nacional de la Historia, 1999), 209–210; Balthasar Ramírez, 301; Armendaris, 133; Robins, *Mercury, Mining and Empire*, 63–64.

and curaca. While this was a means of escaping the mita, it did not necessarily mean they would not work in the mines or smelters. Many of Huancavelica's miners were also hacendados, who saw the value of a large, indebted resident labor pool from which they could draw for their mining activities, or rent them to other miners. Corregidores and curacas also "rebranded" originarios as forasteros so that they would remain locally available as labor and as a market for reparto goods. Similarly, while some clergy were morally opposed to the mita, it also drained a resource, and income, from priests in mita provinces.[35]

The difference between slavery and the mita is tenuous and becomes almost meaningless when one considers the coercive nature of the work and that, for many mitayos, the service directly or indirectly cost them their lives. In these cases, which were not few, their bondage, or its effects, were lifelong. In addition, although they were not technically chattel, they were treated as such, and often rented out by those to whom they were assigned. Finally, while mitayos were to be compensated for their toil, any wages they received were below a subsistence level, which is why they often brought food and family with them to Huancavelica. When they were finally given their wages, they were often only partially paid, and with silver that was so debased that few, including tribute collectors, would accept it.[36]

While mitayos who had completed their term were generally eager to leave Huancavelica, many had little reason to return home. Often, they had missed their harvest or the community lands they previously farmed had been assigned to others, and any livestock left behind had been stolen. Worse, they were quickly assigned new, usually unpaid, tasks by their curaca, corregidor and priest, and once again subject to extortionate fees for reparto goods and religious services. For these reasons, it is not surprising that some mitayos stayed on in Huancavelica to earn a better wage as an alquila.[37]

---

35 Carta de Pedro Camargo, 2; Avisos muy importantes, 3, 5, 8; Representación de Dn. Mathias Chuquimanqui, 1–2; Don Ventura de Santelices y Venero, corregidor de Potosí, sobre la falla de 54 mitayos que anualmente acusa la provincia de Cochabamba con grave perjuicio para la real hacienda. Cochabamba, September 1, 1755 – La Plata, February 27, 1756. ABNB, ALP Minas 127/10, 4–5; Basadre, *El Conde de Lemos*, 140; Lohmann Villena, *Las minas de Huancavelica*, 431; Choque Canqui, 121, 124; Saignes, "Las etnias de Charcas," 49–51, 53. 73 Robins, *Mercury, Mining and Empire*, 49.

36 Memorial del capitán Don Pedro Gutíerrez Calderón, 2; Carta del Virrey del Perú # 38, 3; Expediente promovido por don Nicolás de Sarabia y Mollinedo, 24; Armendaris, 162; Torres de Portugal, 207; Navarra y Rocaful, 164; Velasco, 113; Fonseca, 351; Lohmann Villena, *Las minas de Huancavelica*, 81–82, 175, 232, 386–387, 421; Patiño Paúl Ortíz, 39, 220; Vargas Ugarte, *Historia general del Perú*, Vol 2, 202; Robins, *Mercury, Mining and Empire*, 33, 60–61.

37 Mesia Venegas, 99, 104; Ayans, 35–36, 39–42, Toledo y Leiva, 182; Mendoza y Luna, "Relación del estado del Gobierno," 39; Fernández de Castro, "Carta del Conde de Lemos," 158;

A Dubious Debate

The mita challenged both the law and, for some, their consciences. In 1541, King Charles I had formally, albeit ineffectively, banned forced Indian labor. Subsequent edicts in 1549 and 1551 explicitly banned forced mine work, which had traditionally been assigned to slaves and criminals.[38] This prohibition was reiterated in 1568 by his successor, Philip II, the year he appointed Viceroy Toledo. In 1574, a year after the mita had been instituted, Philip II reversed his position and allowed forced Indian mine work, reasoning that the natives were "indolent by nature."[39] By 1589 he had formally ratified the mita system in Peru.[40]

This monarchical ambivalence on forced Indian labor reflects the broader moral debate which surrounded it. On the one hand there were those like Toledo's collaborator Juan de Matienzo who equated the Indians to animals and explicitly purveyed the Aristotelian notion that they were "born to serve," and that forcing them to toil in the mines was simply "for their own good."[41] Although Matienzo did acknowledge the Indian's physical stamina, he also insisted that "the more strength they have in the body, the less they have in the mind."[42] Such views were commonly held by civil and religious officials alike,

---

Montesinos, 11; Caravantes, Vol. IV, 85; Matinezo, *Gobierno de Perú*, 21–24; Acosta, José de. *De procuranda indorum salute*, 505; Fonseca, 351; Sarmiento de Sotomayor, 87; Lohmann Villena, *Las minas de Huancavelica*, 439; Carrasco, 134; Wiedner, 375; Cobb, *Potosí and Huancavelica*, 63. See also NA, "Pareceres de los Padres," 122; Saignes, "Las etnias de Charcas," 35; Barnadas, *Charcas*, 282.326–27; Lohmann Villena, *Las minas de Huancavelica*, 439; Robins, *Mercury, Mining and Empire*, 149–167.

38   NA, "Pareceres de los Padres de la Compañía de Jesús," 122; Oñate, 141; Juan Sebastián, et al. "Parecer de los PP. de la Compañía de Jesús, Juan Sebastián, Esteban de Avila, Manuel Vásquez, Juan Pérez Menacho y Francisco de Vitoria, dado al Virrey D Luis de Velaso, sobre si es lícito repartir indios a las minas que de nuevo se descubrieren. 1599," in *Pareceres jurídicos en asuntos de indias (1601–1718)*. Rubén Vargas Ugarte, Ed. (Lima: CIP, 1951); 90; Oñate, 141; Vargas Ugarte, *Historia general de Perú*, Vol. II, 207; Robins, *Mercury, Mining and Empire*, 66. For an extended discussion concerning the debates over the mita, see Robins, *Mercury, Mining and Empire*, 62–70.

39   Caravantes, Vol. IV, 47, 213; Castañeda Delgado, "Un capítulo de ética Indiana española," 829–834, 843; Cortés Salinas, "Una polemica en torno a la mita," 131–132; Robins, *Mercury, Mining and Empire*, 66.

40   Cole, *The Potosí Mita*, 20; Robins, *Mercury, Mining and Empire*, 68.

41   Matienzo, *Gobierno de Perú*, 16–19; 80 135; Silvio Zavala, *El servicio personal de los indios en el Perú*. Vol. I (Mexico City: El Colegio de Mexico, 1978), 51; Robins, *Mercury, Mining and Empire*, 62–63.

42   Matienzo, *Gobierno de Perú*, 17.

such as Pedro Gutiérrez Calderón who remarked that "it looks like God created them just for this task."[43]

Similarly, Pedro Camargo, a miner in Huancavelica writing in 1595, averred that "the Indians that are vice ridden understand nothing other than drinking and getting drunk, and once they are drunk they do disgusting things like beasts." He went on to reason that Huancavelica's cinnabar deposits had been

> created by our Lord... for two things: one, so that their richness serves as a source... of a great quantity of mercury to work silver mines... The other is that Our Lord created them so that the Indians...would be occupied in working them because as they are barbarous people... only occupied in their idolatries, drunkenness and other dirty vices... [for] these people to go and work the mines appears to me in the service of God and Your Majesty and for the good of the very Indians...because in the mines they catechize them, make them attend mass and they interact with Spaniards where they become *ladinos* [Hispanicized]
> ... and do not have a place for the idolatries of their lands as they are away from them.[44]

Overall, the deprecation, and dehumanization, of indigenous people and their accomplishments sought to justify their exploitation. Characterizations of Indians as borderline sub-humans reinforced the narrative that the Spaniards were "civilizing" them, while at the same time attenuating any qualms concerning their exploitation. The Inca were likewise vilified, with Juan de Matienzo leading the charge in his *Gobierno del Perú*, a delusional masterpiece of indigenous denigration. In it, he hypocritically characterized the Inca as illegitimate rulers who forced their rule on unwilling subjects, and further condemned them for organizing labor drafts and forcibly relocating communities. He juxtaposed this portrayal with one in which the Indians were joyful under Spanish rule, which he insisted defended their freedoms and protected them from the seizure of their property.[45]

---

43 Memorial del capitán Don Pedro Gutíerrez Calderón, 2. Muñiz, "El Dr. Muñiz de Lima sobre el serujo de los Indios," in "Pedro Muñiz, Dean of Lima and the Indian Labor Question (1603)," in *Hispanic American Historical Review*, K.V. Fox, Vol. IV2, No. 1 (1962), 78; Lizárraga, 201–204; Ulloa, *Noticias* americanas, 329; Molina Martínez, *Antonio de Ulloa*, 85, 329, Patiño Paúl Ortíz, 213; Doña Nieves, 278; Robins, *Mercury, Mining and Empire*, 63–64.

44 Carta de Pedro Camargo, 1.

45 Matienzo, *Gobierno de Perú*, 3, 7, 14–16, 19; Zimmerman, 92, 104–105, 107; Robins, *Mercury, Mining and Empire*, 24–25.

It is ironic that the clergy, who did more than most to expose the abuses and consequences of the mita, would be among the architects of a system that would, directly or indirectly, send thousands to their graves. As early as 1559, the Franciscan, Jesuit and Dominican orders expressed their support for forced mine labor among Indians. In late 1570, Viceroy Toledo established a council to evaluate the mita in terms of its morality, legality and economic impact bringing together clergy, royal officials and legal experts. Although some clergy later retracted their positions, especially as it concerned forced labor in Huancavelica, the council unanimously endorsed the system. A short time later, the Archbishop of Lima, Francisco de Loaiza and the leaders of religious orders began the task of drafting the regulations under which the system would, or rather would not, operate.[46]

Stripped of nuance and casuistry, the Catholic Church maintained a simple and consistent position on the mita. Accepting that the Indians were born to serve, they maintained that the mineral riches of the Americas were a divine gift to Spain that could and should be used to spread the faith and counter heresy.[47] They also recognized the immense hardships and sometime fatal risks associated with the mita, and readily acknowledged that forced mine work was traditionally reserved for convicts, not men who were, technically, free. As Viceroy Chinchón put it in 1630, "it is a most terrible subject as a free, innocent, defenseless, poor and afflicted people are condemned to a high risk of death."[48]

These issues were outweighed, however, by the importance of their task, upon which the common good, realm and Catholicism depended. Moral

---

46   Gaspar de Carvajal, Alonso de la Cerda, y Miguel Adrián, 1; Caravantes, Vol. IV, 52; NA, "Pareceres de los Padres de la Compañía de Jesús," 118; Zimmerman, 263–266; K.V. Fox, "Pedro Muñiz, Dean of Lima and the Indian Labor Question (1603)," in *Hispanic American Historical Review*. Vol. IV2, No. 1 (1962), 70; Rubén Vargas Ugarte, S.J. *Historia general del Perú. Virreinato (1551–1596)*, Vol 11 (Lima: Carlos Milla Batres, 1966), 202–204; Lohmann Villena, *Las minas de Huancavelica*, 99–102; Ramón Ezquerra Abadia, "Problemas de la mita de Potosí en el siglo XVIII," in *La mineria hispana e iberoamericana*. Vol. I. N.A. (Leon, Spain: Catedra de San Isidoro, 1970), 485; Bakewell, *Miners of the Red Mountain*, 64; Robins, *Mercury, Mining and Empire*, 66–67.

47   Caravantes, Vol. IV, 215–216, 220–221; Lohmann Villena, *Las minas de Huancavelica*, 294, 300–301; Casteñeda Delgado, "Un capítulo de ética Indiana española," 883, 866, 881, 916; Fox, 71; Acosta, *Historia natural*, 220, 234; Lewis Hanke, *Bartolomé Arzáns de Orsúa y Vela's History of Potosi* (Providence: Brown University Press, 1965), 7; Fox, 63–65; Patiño Paúl Ortíz, 216; See also Juan de Solorzano Pereira, *De Indiarum Iure*. C. Baciero et al., Eds. (Madrid: Consejo Superior de Investigaciones Científicas, 2001), 541; Robins, *Mercury, Mining and Empire*, 67–68.

48   Gaspar de Carvajal, Alonso de la Cerda, y Miguel Adrián, 1; Carta del Virrey del Perú # 38, 3.

concerns were tempered by the fact that the mitayo's service was to be for a limited time and compensated. In the end, the position of the clergy was that there was "nothing bad" with the mita *per se*, but rather that problems resulted from the manner in which the system operated.[49] Among those beneficiaries in the town of Huancavelica were the clergy, whose parishioners, and servants, included rotating contingents of mitayos and indios de plaza whom they could charge a host of fees during their stay in the town.

The vigorous and prolonged debate over the mita underscores the fact that mitayos were not alone in their familiarity with its abuses, or with the dangers of the mercury mines. Friars, protectores de naturales, governors and even viceroys entered the Santa Bárbara mine, and news of its horrors reached monarchs. In the end expediency triumphed over morality or ethics, with lethal consequences for laborers, both mitayos and wage earners.

Although the mining mita endured throughout the colonial era, there were sporadic initiatives to abolish it. The most concerted vice regal effort was led by Viceroy Pedro Antonio Fernández de Castro, Conde de Lemos. At thirty-five years old, he was remarkably young for a viceroy, and deeply troubled by what he learned about the mita. Following his arrival in Lima in late 1667, he quickly began to try to reform the system in Huancavelica by sending penal labor from Callao, banning night shifts in the mines, prohibiting the gremio from having faltriquera Indians, relieving mita captains of their financial responsibility for missing mitayos, and ensuring more timely dispatches of funds from Lima. Bureaucratic intransigence and miner resistance stifled the measures, prompting him to make an unsuccessful attempt to dismantle the entire mita system. In the process, he implored King Charles II (1661–1700) to support him, flatly telling him "I did not come to the Indies to risk my salvation" and insisting that without regal action the natives would be exterminated.[50] Lemos efforts bore little fruit, and he died in 1672.[51]

---

49   Caravantes, Vol. IV, 215–216; Avendaño, 271; Mesía Venegas, 94–95, 100, 115; Oñate, 142; Carta del Virrey del Perú # 38, 2; Haenke, 138; Acosta, *De procuranda indorum salute*, 515–519, 523, 525, 531, 537; Toledo y Leiva, 180; Casteñeda Delgado, "Un capítulo de ética Indiana española," 879, 883, 886–887, 902, 908, 914–916; Iden, "El tema de las minas," 345; Patiño Paúl Ortíz, 218–219; Bonilla, "1492 y la población indígena," 105; Javier de Ayala, "Estudio preliminar," in *Servidumbres personales de indios*, Javier de Ayla, Ed. (Sevilla: Escuela de Estudios Hispanos-Americanos, 1946), xx, xxv; Fox, 70–71; Perez Tudela, 364, 366; 355–371; Robins, *Mercury, Mining and Empire*, 67–68.

50   Fernández de Castro y Andrade, "Carta del Conde de Lemos a S.M.," 159–162; Basadre, *El Conde de Lemos*, 142–144, 264, 267, 274, 327; Crespo Rojas, "La 'mita' de Potosí," 181; Patiño Paúl Ortíz, 225, 228, 252–253, 255; Cole, 98, 100; Vargas Ugarte, *Historia general del Perú*, Vol II, 206; Robins, *Mercury, Mining and Empire*, 69–70.

51   Basadre, *El Conde de Lemos*, 293–296; Robins, *Mercury, Mining and Empire*, 69.

There were no subsequent efforts to abolish the system until the early eighteenth century. Around 1717, Viceroy Carmine Nicolás Caraciolo, Prince of Santo Buono, proposed shuttering the Santa Bárbara mine, substituting its production with the importation of mercury from Almadén and elsewhere. This idea was, however, quickly spurned by the Council of the Indies given the reliance Spain would have on other nations for the generation of its wealth. Soon after, in 1720, King Philip V made the decision to simply abolish the mita. As with almost all royal edicts, however, it could be suspended by the viceroy pending clarification or to review the feasibility or impact of the measure. Viceroy Caraciolo's successor, Diego Morcillo Rubio de Aúñon, Archbishop of Lima, suspended the measure subsequent to the protestations of the mining gremios in both Huancavelica and Potosí.[52]

Despite this, in 1723 Huancavelica's governor, José de Santiago-Concha y Salvatierra, Marqués de Casa Concha, prepared to substitute convict labor for that of the mita, and sought to reduce the price paid to the gremio for their product. Despite constructing a prison near the entrance to the mine, the epidemic of 1719–1721 had effectively condemned the effort before it began. This devastating event, which in some places had mortality levels approaching fifty percent, also resulted in a labor shortage in both the galleries and smelting operations, and an increase of labor prices. This was the last substantive effort to end the mita, prior to its fleeting abolition on November 9, 1812 by the Spanish Cortés. Following the restoration of Ferdinand VII in late 1813, the system was briefly reinstated before the monarch reversed himself and abolished it by royal edict on April 27, 1820. By this time less than one-hundred Indians were serving the mita in Huancavelica, all of whom came from the region of Chumbivilcas.[53]

Wages and Coercion

The secular decline of mitayos arriving in Huancavelica highlights two interconnected labor-related dynamics: the transformation of the mita from a tax

---

52 Pearce, "Huancavelica 1700–1759," 679; Brown, "Worker's Health," 483; Patiño Paúl Ortiz, 225, 258–259; Pearce, "Huancavelica 1700–1759," 681; Robins, *Mercury, Mining and Empire*, 70.

53 Marqués de Casa Concha, Relación del estado, 24, 27–30, 61; Armendaris, 172; Fernández Alonso, 353, 355–357; Povea Moreno, *Retrato de una decadencia*, 251; Iden, "Entre la retórica y la disuación," 207, 210; Pearce, "Huancavelica 1700–1759," 680; Tandeter, *Coercion and Market*, 32; Brown, "Worker's Health," 483; Carrasco, 223; Patiño Paúl Ortíz, 258–260; Whitaker, 20; Sánchez Albornoz, "Mita, migraciones y pueblos," 31;Basto Girón, 218; Martín José de Mugica, 223; Robins, *Mercury, Mining and Empire*, 70.

paid in labor to one paid in specie, and the prevalent use of wage labor to produce mercury. By the mid-seventeenth century, alquilas made up half if not more of the workforce, earning up to a peso and a half each day, a good wage for the region.[54] By 1750, only Angaraes, Chumbivilcas and Cotabamba continued to send men, with the other mita provinces remitting commutation payments to gremio members which amounted to around 40,000 pesos. Although these were theoretically destined to hire alquilas, some miners simply pocketed the funds and forsook production. After 1779, when the crown abolished the gremio and took over direct operation of the mine, the revenue from the "mita tax" accrued to the treasury. By 1790 the mitayo and base alquila wage were at parity, underscoring the eroding difference between wage and mita labor.[55]

The rise of a wage labor force in Huancavelica raises questions about what were widely reported as atrocious conditions in the mines. If the working environment did not preclude the emergence of wage laborers, then, the argument goes, things could not have been as bad for the mitayos as many have suggested.[56] Much of the rise of the alquila workforce occurred, however, after the opening of the Bethlehem adit in 1642 and the consequent improvements in ventilation and access to the mine. By this time, most of the epidemics which had ravaged the region had passed, leaving in their wake an increased demand for labor.[57]

---

54   Pedro Fernández de Castro y Andrade, Conde de Lemos, "Advertencias que hace el Conde de Lemos a la relación del estado del reino que le entregó la Real Audiencia de Lima del tiempo que gobernó en vacante de virrey que fue de año y más de ocho meses, dirigida a la reina nuestra señora en el real y supremo consejo de Indias," in *Los virreyes españoles en America durante el gobierno de la casa de Austria. Perú*. Vol. IV. Lewis Hanke and Celso Rodríguez, Eds. (Madrid: IMNASA, 1978), 254. See also Fernández de Castro, "El Conde de Lemos da cuenta," 271; Haenke, 136; Lohmann Villena, *Las minas de Huancavelica*, 242, 253, 330–333, 353, 382; Brown, "Worker's Health," 485; Robins, *Mercury, Mining and Empire*, 62.

55   Expediente promovido por don Nicolás de Sarabia y Mollinedo, 8–9, 14; Haenke, 136; N.A., "Memoria sobre la mina de azogue de Huancavelica," 90, 113; Mugica, 224; Molina Martínez, *Antonio de Ulloa*, 82; Fisher, *Silver Mines and Silver Miners*, 11, 92; Povea Moreno, "Entre la retórica y la disuación," 202; Iden, *Retrato de una* decadencia, 190; Melissa Dell, "Persistent Effects of Peru's Mining Mita." In *Essays in Economic Development and Political Economy*. (Diss. Massachusetts Institute of Technology, 2012), 20; Molina, Martínez, 82; Robins, *Mercury, Mining and Empire*, 62; Brown, "Worker's Health," 485.

56   Bradby, 228–230.

57   Emetherio Ramírez de Arellano, 2–3; Fernández de Castro, "Advertencias que hace el Conde de Lemos," 254. See also Haenke, 136; Fernández de Castro, "El Conde de Lemos da cuenta," 271. Haenke, 136; Lohmann Villena, *Las minas de Huancavelica*, 242, 253, 330–333,

Further downplaying toxic conditions in the mines and smelters were comforting statements from governors and viceroys minimizing, or denying, the existence of mercury poisoning. For example, after Viceroy Mendoza entered the mine in 1608, he reported that "the present condition of the mine is not of more work, nor of more danger, in health and life, than other mines."[58] Despite this, he did concede that "every day there are accidents and rockslides" and related how "I encountered some...who were sick... with fevers and other illnesses."[59] It is ironic that the viceroy would minimize the severe conditions in the mine, as, after climbing the ladders to enter and exit the mine, he had such severe cramps that he was unable to walk for four days. Similarly, Governor Ulloa reported that "no longer are there tragedies, nor deaths in the mine nor in the smelting places."[60]

This denialist line has been embraced in recent work arguing that mine and smelter labor in Huancavelica was not as dangerous as many suggest given the longer tenure of wage laborers there relative to mitayos and the prevalence of forasteros in the region.[61] Such a position does not, however, account for several factors. One is that alquilas could largely choose the work they did, eschewing exceptionally dangerous tasks and preferring to work as barreteros or on the surface as ore sorters or transporters. As a result, they were not exposed to the same dangers, quotas or abuses that mitayos suffered. In contrast, mitayos did the most dangerous tasks, were exposed to dust and vapor from cinnabar for longer uninterrupted periods, were subject to more abuse and probably less well fed, all of which would have a cumulative effect on their health.

353, 382; Brown, "Worker's Health," 485; Cook, *Demographic Collapse*, 69, 208; Robins, *Mercury, Mining and Empire*, 62.

58   Juan de Mendoza y Luna, Marqués de Montesclaros, "Carta del Virrey Marqués de Montes Claros a S.M. informando extensamente sobre las minas de Guancavelica, en virtud de la comunicación y conferencias que sobre el asunto tuvo con su antecesor en aquel gobierno D. Luis de Velasco." In "Cuatro cartas del marque´s de Montesclaros referentes a la mina de Huancavelica." Manuel Moreyra y Paz Soldán, ed. *Revista Histórica* 18 (1949), 94.

59   Juan de Mendoza y Luna, Marqués de Montesclaros, "Carta del Virrey Marqués de Montes Claros a S.M. en materia de Real Hacienda, cantidades que se envian de todo género de hacienda; ruina del Cerro de Guancavelica para lo que pide socorro de azogues; estado que tiene el edificio de la iglesia de Lima, tiempo en que podrá terminar y lo que ha costado a S.M." In "Cuatro cartas del Marqués de Montesclaros referentes a la mina de Huancavelica." Manuel Moreyra y Paz Soldán, ed. *Revista Histórica* 18 (1949), 89; Iden, "Carta del Virrey Marqués de Montes Claros a S.M. informando extensamente sobre las minas de Guancavelica," 94.

60   Lohmann Villena, *Las minas de Huancavelica*, 22, 229; Scribner, 34–35, 41; Molina Martínez, *Antonio de Ulloa*, 88.

61   Bradby, 228, 230–232.

In addition, when a worker, whether an alquila or mitayo, became ill, they usually endeavored to return home or left to warmer climes, and would not be listed as perishing in Huancavelica.[62] Moreover, if the mita was not so widely feared or despised by those who were subject to it, then there would not have been wholesale migration to avoid the service, nor would an institutionalized commutation system have emerged.

The minimalist position begs the question as to just how free were "free" wage laborers? With the conquest, several forces compelled Indians into a cash economy and made subsistence production less viable. One was the decline of the vertical archipelago system due to the demographic decline and forced resettlement into reducciónes. Another was dispossession of community lands as a result of the composiciónes de tierras, and the loss of access to what remained as people fled the mita to become forasteros. Put differently, the more circumscribed Indians' relationship with their environment was, the more they were reliant on a cash economy. Other forces were also at work, such as the need to pay tribute, reparto debts, religious fees and mita commutations in specie. In the region of Huancavelica, curacas also rented out people from their community to serve as alquilas, generating funds for the community to pay commutation fees and other debts.[63]

Even debt peons on haciendas, who were otherwise to some degree protected from the reparto, tribute and the mita, were also dispatched by their patrons to work in their or other's mines. The result was that many of the alquilas in Huancavelica were there to pay debts or meet other financial exactions which had been forced upon them by corregidors, priests, curacas, or contracted during their mita term. Within a context of such multi-layered economic, political and cultural coercion, free will is inescapably compromised. Thus, while there was indeed a sizable wage labor force in Huancavelica, given the circumstances in which they made their decisions, it can hardly be considered "free."[64]

Mercury production transformed man's relationship to the environment in Huancavelica at many levels, both physically and figuratively. While pre-Hispanic cinnabar mining was largely limited to surface extraction and the production of vermilion, the thirst of the colonial economy for quicksilver seemed

---

62   Memorial del capitán Don Pedro Gutíerrez Calderón, 1; Scribner, 39–40.

63   Capoche, 174; Carrió de la Vandera, 162; Zulawski, "Wages, Ore Sharing and Peasant Agriculture," 415; Iden, *They Eat From Their Labor*, 112; Patiño Paúl Ortíz, 245, 247; Brown, "Worker's Health," 470.

64   Favre, "Evolución y situación de las haciendas," 240; Iden, "La industria minera de Huancavelica en la década de 1960," in *Boletín de Lima*, No. 161 (2010); Barnadas, *Charcas*, 282; Fisher, *Silver Mines and Silver Miners*, 10.

inexhaustible and ultimately depended on shaft-based production. Those who were involved in the mining and smelting of cinnabar were coerced into doing so, either directly through the mita or their curacas or hacendado masters, or indirectly as a result of the exactions levied by colonial or ecclesiastical authorities. Just as the nature of cinnabar extraction changed after the conquest, so too did the mita. Although the Inca and previous groups employed forced labor, it tended to be for shorter periods, was generally less onerous, and often provided public benefits in the form or roads, bridges and granaries. Under the Spanish, the concept of public good included that which benefitted Spaniards and creoles personally. Furthermore, the length and terms of mita service were considerably harsher, and the protections of the law did little to alleviate the Indians. Had people not abhorred mita service, there would not have been the mass migration to avoid it by becoming classified as a forastero, nor would the commutation payments have been formalized in Huancavelica's subject provinces.

The mita system caused some soul searching, especially among clergy, but they came up empty-handed. As a whole, they were clearly discomfited by the coercive nature of the mita system, especially as it applied to mine work given its association with slavery and criminal punishment. In the end, such misgivings were outweighed by the self-serving belief that the mineral deposits in the Americas were divinely placed there for the benefit of the Spanish, who had an obligation to use them to spread and defend Catholicism. They lamented that this was at the expense of the Indians, but rather than condemn the system in which they had played a leading role in designing, they criticized its operation.

Unfortunately for the region's indigenous inhabitants, they had few options, and little to come home to after a stint in Huancavelica. Often they had missed the harvest, lost livestock and access to land, owed tribute payments, and in many cases were physically incapacitated. Those who were able were soon pressed into tasks by their curaca, had reparto goods foisted upon them by their corregidor and had services and fees due to their priest. As a result, many mitayos never returned home, and some even stayed on in Huancavelica as alquilas.[65] Colonial authorities were hard pressed to maintain production

---

65   Mesía Venegas, 120; "Informe del corregidor don Juan Medrano Navarrete sobre la situación de los pueblos que componen la provincia de Pacajes," in *Potosi: La version aymara de un mito europeo. La mineria y sus efectos en las sociedades andinas del siglo XVII* (*La Provincia de Pacajes*), Teresa Cañedo-Arguelles Fábrega (Madrid: Editorial Catriel. 1993), 109; Fisher, *Silver Mines and Silver Miners*, 10; Bakewell, *Miners of the Red Mountain*, 111–113; Cole, 26–27, 49; Saignes, "Las etnias de Charcas," 35; Brown, "Workers' Health," 470; Robins, *Mercury, Mining and Empire*, 35.

in the context of constantly declining numbers of mitayos, deteriorating ore quality, and, in the early eighteenth century, another devastating epidemic of influenza and smallpox. With the rise of the Bourbon family to the Spanish throne in 1700, productive and administrative issues in Huancavelica received increasing attention, ultimately affecting the environment and people of the region.

CHAPTER 4

The Bourbon Era: Reform and Resentment

The coronation of Philip V in late 1700 as Spain's first Bourbon monarch heralded a time of considerable political, economic and social change in the Americas, and in Huancavelica. Overall, Bourbon initiatives in Huancavelica were focused on administrative and fiscal efficiency and increasing mercury production Central to this effort were the introduction of improved furnaces, the importation of tools and technical expertise from Spain, and the elimination of the quinto on mercury production in 1760, all of which resulted in an increase of production through most of the eighteenth century.[1] Given better oversight and administration, some of the apparently increased production was in fact mercury that would otherwise have been diverted to the black market but was instead registered. By ensuring the consistent provision of royal funds to the town, miners had less reason to illicitly sell mercury to silver refiners or merchants at a discount. By the 1790s, however, Huancavelica's production was faltering. This reflected a general decline in ore quality, the failure of the crown's effort of direct operation of the mines which began in 1782, and a major cave-in in 1786 which shut part of the mine for what remained of the colonial era and beyond.[2]

Bourbon innovations got off to a slow start, however, as the turmoil of the War of Succession (1701–1714) reduced what royal oversight there was in the town. This, combined with the expansion of French merchant activity on the Pacific coast and the increased sales of public positions, all facilitated illicit trade. Much of this was paid for in unminted, and contraband, silver, which in the region of Huancavelica was often refined with untaxed mercury. The numbers of mitayos arriving in Huancavelica, and official production levels of quicksilver, declined, with the smelters rendering an average of 152 metric tons a year between 1700 and 1715. Meanwhile, the amount of gremio debt, and the number of absentee gremio members, increased as more miners accepted

---

1   Fuentes Bajo, 97; Sánchez Gómez, "La tecnica en la producción," 166; Whitaker, 68–71; Matilla Tascón, Vol. II, 286.

2   Pearce, "Huancavelica 1700–1759," 694, 700–701; Haenke, 265; Crosnier, 47–48; Patiño Paúl Ortíz, 89, 91–92, 134; Carrasco, 243, 247; Lohmann Villena, *Las minas de Huancavelica*, 361; Roel Pineda, 117; Whitaker, 57–62. 65–66; Brown, "La crisis financiera," 364; Iden, "La distribución del mercurio," 157; Fisher, *Silver Mines and Silver Miners*, 21; Robins, *Mercury, Mining and Empire*, 56, 170.

mita commutation payments and used them as personal income as opposed to investing in production. Other gremio members rented out their mines to tenants whose desire for profit led to especially abusive treatment of mitayos and even more extraction from the mine's structural supports. Huancavelica was, however, still able to provision the Peruvian viceroyalty, and occasionally dispatch mercury to New Spain. For example, in the 1720s, the Andes consumed around 161 metric tons of mercury annually, which is about what Huancavelica produced in 1723 and 1724.[3]

In 1723, the fifty-six year old creole audiencia member José de Santiago-Concha y Salvatierra, Marqués de Casa Concha, became governor of Huancavelica. While royal officials often had contradictory instructions, his were especially so. He was tasked with increasing production, decreasing the price of mercury paid to the gremio, abolishing the mita and stamping out the contraband trade of mercury. The conflicting nature of these objectives ultimately undermined his efforts, as did the epidemic of 1719–1721, which led to a severe labor shortage, and increased wages, in Huancavelica and elsewhere. In an effort to maintain production, Casa Concha sought unsuccessfully to introduce convict labor in the mines and, pending the end of the draft labor system, prohibited the payment of silver in lieu of mita service.[4]

Likewise, the plan to reduce the amount of money paid for mercury to the gremio, from fifty-eight pesos to forty per *quintal*, or forty-five kilograms, was unsuccessful. The gremio objected strongly, and even engaged in a work slowdown. Not only were they protesting the reduction of the price of mercury, but also the prospect of the abolition of the mita, which would require them to pay seven to eight reales a day to wage laborers, instead of the three or four which mitayos received at that time. In the end, the mita, and mercury price, survived, as Casa Concha prioritized production over price and labor reforms, surmising that the price reduction would result in a spike in contraband. Despite this, Casa Concha did have some successes. By utilizing his authority to draw funds from other treasury offices in the region, as opposed to Lima, he was in a better position to disburse payments when they were needed, thereby discouraging contraband. He also improved the safety conditions inside the mine through better maintenance, and expanded the hospital while subjecting its finances to a degree of civilian oversight.[5]

---

3 Pearce, "Huancavelica 1700–1759," 671–674, 694; Patiño Paúl Ortíz, 130; Fernández Alonso, 359–360; Robins, *Mercury, Mining and Empire*, 95.

4 Fernández Alonso, 349–353, 355–356, 370–371; Povea Moreno, *Retrato de una decadencia*, 251–253.

5 Fernández Alonso, 357–358, 364, 366–369; Povea Moreno, *Retrato de una decadencia*, 268.

If anything, Casa Concha's tenure highlighted how difficult any substantive change to the status quo in Huancavelica was going to be. In an effort to increase the accountability of the town's administration, and in the broader context of Bourbon efforts to centralize power, beginning in 1736 governors were appointed directly by the king. As a result, the system of appointing oidores from Lima's audiencia was abandoned, and their interests consequently harmed. The first corregidor so chosen was Jerónimo de Sola y Fuente, who would serve in Huancavelica for a dozen years. Often referred to as the "Restorer of the Mine," Sola y Fuente epitomized the new breed of modernizing, efficiency-minded, Bourbon administrators who would take the reins in Huancavelica.

Among his assignments was the abolition of the gremio and the institution of direct royal operation of the mines. After studying the issue, however, Sola y Fuente determined that the royal interest, and mercury production, would not be well-served by such measures. He reasoned that the crown would lose the revenue from the quinto and incur considerable expenses in operating the mine. Moreover, he would have to contend with vehement gremio opposition and the risks of sabotage and declining production during the shift to direct operation. Instead of abolishing the gremio, he negotiated new terms with them which were expressed in the asiento of 1744. Sola y Fuente's concerns about the viability a royal takeover of mercury production were well-founded. When the crown finally implemented the policy in 1782, production declined to such a point that it could not supply even half of Peru's needs. Beyond that, not only did direct operation not generate tax revenue, but the crown incurred a loss of thirty-eight pesos per quintal of mercury produced.[6]

Assisting Sola y Fuente in his charge of increasing production were a group of Spanish mining experts who had worked in Almadén. With their assistance, he was able to increase mercury production, which ranged under his administration from 596 to 826 metric tons per year. Beyond expertise, this increased production resulted from re-encountering the main cinnabar deposits in the Santa Bárbara mine, the use of gunpowder for dislodging ore, and from refining lower quality ore and tailings. To the relief of local communities, and those who exploited them, he also reduced mita demands. Sola y Fuente also stripped absentee miners of their mita allotments and maintained a steady flow of funds for the gremio. As a result of the work of Casa Concha and Sola y Fuente, mercury production increased in the first half of the eighteenth century, miners had better access to credit and prompt payment for the

---

6 Abrines, 108; Pearce, "Huancavelica 1700–1759," 683–684, 690; Carrasco, 229; Whitaker, 24–26, 62, 71–72; Robins, *Mercury, Mining and Empire*, 31.

quicksilver they produced, and as a result contraband trade of mercury was attenuated.[7]

Philip V's successor, Ferdinand VI, likewise sought to strengthen royal control and increase mercury production in Huancavelica. With this in mind, he named Antonio de Ulloa as governor, selecting him for his scientific knowledge and integrity. Ulloa already had extensive experience in Latin America when he arrived in Huancavelica in November, 1758. His time there, however, was infelicitous, or, as he described it, a "purgatory."[8] The source of many of his troubles was his effort to stem the endemic corruption in the town, precipitating fractious relations with the gremio amid Ulloa's threats to dissolve it. Bringing tensions to a head was his arrest of two alcaldes/veedors and another official who had engaged in the time-honored practice of mining supports and selling the rich ore they yielded. Further undermining his relations with the gremio was his decision to arrest the leaders of, and abolish, the Spanish militia after their refusal to parade with that of mestizos. Ulloa was ultimately forced by another rival, Viceroy Amat y Junient, not only to release the militia leaders, but also to thank them for their service. This did little to heal the rift, nor did his experiment with direct operation of the mine through the establishment of the *Minería del Rey*, or king's mining company. Not only did it compete with the mining guild, but also served as a reminder that they were not irreplaceable.[9]

Ulloa did not just have problems in Huancavelica, but also in Lima. Since he was appointed by the king, and not the viceroy, he refused to pay the customary bribe of 10,000 to 12,000 pesos to the viceroy for the position. For once, the Spanish viceroy and largely creole audiencia in Lima had something in common. Although the audiencia also resented Ulloa because his appointment disrupted the previous system of rotating appointments, and concomitant opportunities for graft, they did not lament the diminution of vice regal power that Ulloa's appointment entailed.[10] Despite the conflicts and obstacles which he faced, Ulloa's time in Huancavelica was not unsuccessful. Among his innovations was the use of iron supports to increase the structural integrity of the mines, as opposed to the masonry construction which had been used until then. He also improved drainage inside the mine and demanded more effective oversight from veedores. In addition, he began the construction of a new vent shaft, and like many Bourbon administrators, focused on public

---

7   Carrasco, 228–29; Pearce, "Huancavelica 1700–1759," 683, 685, 688–689, 690–693, 701–702; Whitaker, 24; Fernández y Alonso, 358–359, 368–369.

8   Abrines, 109, Whitaker, 34; Molina Martínez, *Antonio de Ulloa*, 20, 38; Whitaker, 32, 42.

9   Whitaker, 34, 39–42; Abrines, 110; Molina Martínez, *Antonio de Ulloa*, 43–44; 69, 74, 122.

10   Whitaker, 42–43.

works, such as improving the water supply to the town, paving several streets and constructing a bridge across the Ichu River. When Ulloa's long-requested order arrived calling him to Havana in 1764, Viceroy Amat, the audiencia and the gremio were delighted to see him go.[11]

Casa Concha, Sola y Fuente and Ulloa embodied the Bourbon focus on improved public administration, increasing revenues, suppressing contraband and constructing public works. Among the manifestations of such policies under King Charles III (1716–1788) was the division of the Peruvian Viceroyalty into that of Peru and Río de la Plata in 1776, the latter with its seat in Buenos Aires. Not only did this alter trade routes in the region away from Lima, but it also affected the allocation of mercury. Huancavelica's production would henceforth be almost exclusively dedicated to supplying Castrovirreyna, Huantajaya, Cerro de Pasco, Hualgayoc, San Antonio de Esquilache and other Peruvian mining centers, while Upper Peruvian silver mines were increasingly supplied with quicksilver from Almadén.[12]

In Huancavelica, despite improvements in production, the crown remained dissatisfied with the ability of the decrepit and corrupt gremio to satisfy the region's needs for quicksilver. This was especially the case with Charles III's *visitador*, or special representative with extensive powers, José Antonio de Areche. He had been sent to Peru in 1777 to enforce administrative and other reforms with the goal of increasing revenue and strengthening royal power. Instead of trying to fix the dysfunctional system of asientos with the gremio which had prevailed for almost two centuries, Areche was determined to dismantle it. In May, 1779, he granted Nicolás de Saravía y Moliendo the exclusive privilege of extracting cinnabar and providing mercury to the crown. Saravía promised to construct 100 new smelters and produce 6,000 quintals of mercury annually, which was to be exclusively sold to the treasury at about two-thirds of the previous price. The only consolation for the now extinct gremio was that Saravía had been drawn from their ranks. He had hardly begun his work when he died of pneumonia in December, 1780, although he did succeed in producing 10,300 quintals during this time. As so often, however, this was accomplished through the now inveterate practice of excavating the cinnabar-rich estribos and puentes in the mines. Following Saravía's death, in 1782 the mines

11   Umlauff, 14; Abrines, 110; Whitaker, 39; Molina Martínez, *Antonio de Ulloa*, 25, 67–69, 71, 75 154, 195.

12   Expediente sobre la postura para la conducción de azogue de esta Caja de Potosí a las demas del Virreinato. Potosí, 1781. CNMAH, CGI/M-64/26, 1; Patiño Paúl Ortíz, 89; Basadre, *El Conde de Lemos y su tiempo*, 158; Robins, *Mercury, Mining and Empire*, 71.

came under direct royal administration until independence, and production continued its secular decline.[13]

In 1784, the town's first *intendant*, or governor which replaced the position of corregidor, Fernando Márquez de la Plata, presided over an initial increase of production from a paltry seventy metric tons in 1784 to 206 in 1785. Like Saravía, Márquez' director of mines, Francisco Marroquín, accomplished this by extracting from the structural pillars in the mines, resulting in a massive cave-in on September 25, 1786. Not only did hundreds of workers perish in the rubble, but much of the Bethlehem adit, along with the Comedio and Cochapata zones of the mine, were flooded. Intendant Márquez tried to cover up the reckless mining practices by claiming that the damage had been caused by an earthquake. Viceroy Teodoro de Croix was unconvinced, however, and fired Márquez and jailed Marroquín until his death sixteen years later.[14] In many ways, this cave-in ended Huancavelica's reign as the "soul or spirit" of the Andean silver mills, with production in 1787 being half that of the previous year.[15] Much of that came from the Trinidad mine, and after the 1790 discovery of the Sillacasa deposit also helped to compensate for the production lost from the Santa Bárbara mine.[16]

The 1786 collapse underscored the dialectical relationship between risk and production in the mines. Seeking to overcome this was the scientific

---

13   Expediente promovido por don Nicolás de Sarabia y Mollinedo asentista de la Real mina de azogue, 2–26; Haenke, 266; Whitaker, 57–62, 65–66; Kendall Brown,"La distribución del mercurio a finales del periodo colonial, y los trastornos provocados por la independencia hispanoamericana," in *Minería colonial Latinoamericana*, Dolores Avila, Inés Herrera and Rina Ortíz, Eds. (Mexico City: Instituto Nacional de Antropología e Historia, 1992), 157; Fisher, *Silver Mines and Silver Miners*, 21; Patiño Paúl Ortíz, 90–91; Carrasco 243; Strauss, 561; Lang, 215–217; Vicente Palacio Atard, "El asiento de la mina de Huancavelica en 1779," in *Revista de Indias*, Vol. v (1944), 621–628; Robins, *Mercury, Mining and Empire*, 170.

14   Haenke, 265; Crosnier, 47–48; Patiño Paúl Ortíz, 89, 91–92; Carrasco, 243, 247; Robins, *Mercury, Mining and Empire*, 56. For a discussion of the Intendant system in Huancavelica, see Isabel M. Povea Moreno, *Retrato de una decadencia: régimen laboral y sistema de explotación en Huancavelica, 1784–1814*. Diss. (Granada: University of Granada, 2012), 52–61. On the drawn-out efforts of Pedro Subiela to map the mine and design repairs to it, see Kendall Brown, "El ingeniero Pedro Subiela y el desarrollo tecnológico en las minas de Huancavelica (1786–1821)," in *Histórica*. Vol. IIIo, No.1 (July, 2006): 165–184.

15   Llano Zapata, 225; Carrasco, 247.

16   Haenke, 265; Arana, 11; Rivero y Ustariz, *Memoria Sobre El Rico Mineral De Azogue*, 49; Strauss, 562; Contreras and Díaz, 8; Povea Moreno, "Los buscones de metal,"116; Mervyn Lang, "El derrumbe de Huancavelica en 1786. Fracaso de una reforma bourbónica," in *Histórica*, Vol. Io, No. 2 (Dec., 1986), 21–22; Contreras and Díaz, 7; Robins, *Mercury, Mining and Empire*, 56.

expedition led by Thaddeus von Nordenflicht, which arrived in the town in December, 1790. Nordenflicht had previously spent time in Potosí where he and his team demonstrated and urged the adoption of reverberating ovens for silver processing, with scant success. In Huancavelica, their task was to develop a plan to recover from the 1786 cave-in and otherwise modernize production. As a result of his month-long visit there, Nordenflicht called for new smelters be constructed near the pithead to reduce transportation costs, and proposed revisions to the mining code. More importantly, he urged the construction of a central shaft to increase ventilation and enable men, tools, materials and ore to be lowered into or hoisted from the mine with mule-operated winches.[17] Nordenflicht believed that the central shaft would also enable greater oversight and reduce fraud in the mines, which he referred to as a "building of thieves."[18] In the end, his reforms were not adopted, partly due to concerns about declining ore quality in the mines and, as in Potosí, the costs of his proposals and his arrogance.[19]

Pallequeo and Production

Nordenflicht's visit was not without consequence, however, for if nothing else it showed the scale of investment necessary to modernize the mines. This, in turn, prompted royal officials to consider the unthinkable: largely unregulated mining and refining, subject to the crown monopoly on the purchase of mercury. The result was the legitimization of a system known as *pallequeo*, which consisted of informal, usually surface, mining or refining of tailings conducted by *humanchis*, who were generally indigenous miners and refiners. This was the fifth system of mining employed in the region, following the original ownership claims, the gremio system, a sole contractor, and direct operation by the crown.[20]

---

17   Povea Moreno, *Retrato de una decadencia*, 164; Kendall Brown, "Nordenflicht, Thaddeus von. Tratado del arreglo y reforma que conviene introducir en la minería del reino del Perú para su prosperidad, conforme al sistema y práctica de las naciones de Europa más versadas en este ramo, presentado de oficio al superior gobierno de estos reinos por el barón de Nordenflicht. Estudio preliminar de José Ignacio López Soria," in *Histórica* 31.1 (July, 2007): 213. For an extended discussion of the Nordenflicht expedition, see Povea Moreno, *Retrato de una decadencia*, 157–177 and Whitaker, 68–71.

18   Brown, "Nordenflicht," 213.

19   Whitaker, 71.

20   Povea Moreno, "Los buscones de metal," 109–111, 121–123; Berry and Singewald, 24; Patiño Paúl Ortíz, 130.

In 1792, Intendant Conde Ruíz de Castilla allowed all people, including Indians, the privilege of working in part of the Santa Bárbara mount and also to register and work cinnabar mines for a period of thirty years. This latter provision, however, only applied to mines located fifty kilometers from Huancavelica and required the beneficiary to bring the mercury on a weekly basis to the treasury office in Huancavelica. More established miners opposed the system, fearing that that it would reduce their labor supply and lead to an increase in unregistered production. Other misgivings included concerns that smelter operators would collude with pallequeo miners to only partially refine ore so that the pallequeo workers could re-refine it, and that the system would discourage underground work.[21]

Despite the risks, by 1794 officials deemed the experiment of legal pallequeo a success as production had increased fifty percent. By 1795, there were about 130 *palleaquedores*, or humanchis, working eleven sites in the vicinity of Huancavelica. It was not just the Crown that benefitted through cost savings, so too did workers, who had fewer health risks from surface extraction and reprocessing tailings than with shaft mining. Despite the productive potential of pallequeo, humanchis generally lacked capital and consequently engaged in small scale, often family-based, production while remaining dependent on Spanish and Creole smelter owners. Nevertheless, the rise of pallequeo entailed the rise of the legal independent Indian miner, although their ranks also included mestizos, creoles and even Spanish.[22]

The importance of humanchis only grew with time. With the upheaval of the wars for independence, the abolition of the mercury monopoly by the Spanish Regency on January 6, 1811 and the transitory abolition of the mita in 1812, Spanish miners increasingly withdrew from production, and Huancavelica itself. The humanchis took up the slack, and pallequeo would outlast the colony, and be responsible for almost all mercury production throughout the nineteenth century.[23]

---

21 Whitaker, 72–75, 81; Brown, "La distribución del mercurio," 160; Povea Moreno, *Retrato de una decadencia*, 122; Iden, "Los buscones de metal," 115, 121, 125–127; Robins, *Mercury, Mining and Empire*, 170.

22 Povea Moreno, "Los buscones de metal," 121–123, 130–132, 135; Iden, *Retrato de una decadencia*, 261; Contreras and Díaz, 10; Whitaker, 73. For an extended discussion of pallequeo, see Povea Moreno, *Retrato de una decadencia*, 89–125.

23 Whitaker, 73–74, 90, 92; Arana, 11, 16; Povea Moreno, "Los buscones de metal," 109, 131–132, 135–136; Contreas and Díaz, 3, 23; Palacio Atard, 629; Brown, "Worker's Health," 483; Carrasco, 223; Patiño Paúl Ortíz, 258–260; Whitaker, 20; Povea Moreno, "Entre la retórica y la disuación," 207, 210; Sánchez Albornoz, "Mita, migraciones y pueblos," 31; Basto Girón, 218; "Robins, *Mercury, Mining and Empire*," 70.

The Colonial City

Colonial Huancavelica's growth and prosperity were a function of the productivity of its cinnabar mines. While increased production improved the
economic fortune of the town, it compounded its environmental misfortune.
When Viceroy Toledo upgraded the status of the town to that of a *villa*, or municipality, in 1581 it was freed from the jurisdiction of Guamanga and endowed
with its own governor with authority over a six league radius. At this point, the
town had 170 modest houses concentrated in the center of town, most constructed with local travertine around a patio and corral with straw roofs. The
San Antonio church on the plaza was almost completed, and nearby was the
chronically underfunded and understaffed hospital. Residents obtained drinking water from springs and also the Sigisichaca River, which in the national
period was renamed the Ichu River.[24]

During the first few decades of the 1600s, the town's adult Spanish and creole population rarely exceeded 400, and it had a total population of only a
couple of thousand people. Among its residents were slaves of African descent,
who in this isolated region could sell for over 500 pesos, a substantial amount.
For example, in 1588, of the 253 slaves and ten free people of color living in the
town, most were dedicated to producing various ceramic vessels and tubes for
mercury refining, while others served as blacksmiths, domestic servants and
overseers. During this period, houses increasingly had tile roofs, although the
region's poor quality clay meant that they required frequent replacement. The
region's harsh climate and lack of comforts resulted in a heavily male population who generally maintained their primary home, and families, elsewhere.[25]
Further limiting the population there was the seasonal nature of mercury production, which, until around 1700, largely shut down during the rains, which
lasted from January through May. Although refining had been seasonal, extraction was continuous, especially in the most productive mines such as that of
Santa Bárbara, Santa Inés, Inés de Robles, Santa Isabella, Juan García, Correa
de Silva and Mina Nueva. The declining numbers of mitayos do not appear to
have had much of an impact on the town's population as alquilas took their

---

24   Cantos de Andrade, 307–308; Montesinos, Vol. II, 81–82; Carrasco, 104, 116; Patiño Paúl
      Ortíz, 33–34, 38, 63; Lohmann Villena, *Las minas de Huancavelica*, 116; Contreras, *La ciu-
      dad del mercurio*, 72, 77; Robins, *Mercury, Mining and Empire*, 31.
25   Cantos de Andrade, 305; Carrasco, 104, 127; Reyes Flores, 69; Contreras, *La ciudad del
      mercurio*, 42, 51, 73; Patiño Paúl Ortíz, 34–35, 39, 44; Robins, *Mercury, Mining and
      Empire*, 29.

place. It would not be until the 1650s that Spanish and creole families increasingly took up residence there.[26]

The demographics would have been much different had a planned Indian uprising been successful in 1667. Led by Indians in the San Cristóbal and Ascención neighborhoods, muleteers connected the plotters to conspirators in Lima. Although it was foiled, the plot had both Indian and Negro participation and had as its objective the elimination of the town's white and mestizo population. Slightly more successful was a revolt against the mita in the district of Moya in 1756, which, although quickly suppressed, did succeed in focusing the attention of civil authorities on the conditions in which miners worked.[27]

Writing in the early 1600s, an anonymous chronicler described the town as having

> many Indians that work in the mines... [the town] never lacks merchandise and other people who come to trade in the town, as it is rich... there is raised in the fields and high mountains that are nearby many cattle and sheep from which they make excellent butter and much cheese, and from their meat they make good dried meat. It has much sugar that they get in the deep ravines through which passes the Marañon River, which never freezes. And there are very good farms in those ravines and many fruits.[28]

By 1700, about 500 Spaniards and creoles lived in the town, joined by 520 Indian heads of households, which suggests a population of at least 5,000, exclusive of mestizos and slaves. There was vibrant commerce on the main plaza, some streets had been cobblestoned, and two stone bridges connected the town with the Indian districts on the other side of the river. Seven churches served the parishioners, a remarkable number for such a small town. Huancavelica's population fluctuated considerably over the years. For example, after a decline in the early 1700s, the population increased, along with mercury production, after 1725 to reach about 10,000 in 1750, and around 14,000 in the

---

26   Caravantes, Vol. IV, 134; Haenke, 136; Contreras, *La ciudad del mercurio*, 43; Paúl Ortíz, 15, 35; Contreras, *La ciudad del mercurio*, 45–46; Brown, "Worker's Health," 483; Robins, *Mercury, Mining and Empire*, 30, 54.

27   Government of Peru. *Almanaque De Huancavelica*, 19; Contreras, *La ciudad del mercurio*, 70–71; Carrasco, 188; Patiño Paúl Ortíz, 79–80, 86; Pease, 308, 312. Concerning another planned rebellion in 1611, see Montesinos, 192–193, and a messianic movement in 1811 in Lircay, see Pease, 345–378.

28   Lewin, 83.

1760s.[29] By 1784, however, the town had only 5,472 residents, with 3,000 living in the indigenous San Cristóbal neighborhood, 1,320 in Ascención, 720 in Santa Ana and 432 in the village of Santa Bárbara at the pithead. Over the coming years, the population would increase, again reaching about 10,000 in 1812, and declining to about 5,900 in 1817. Throughout these changes, the ethnic composition remained stable, with roughly three quarters of the population being Indian, fourteen percent mestizo, ten percent Spanish or creole, and one percent Negro.[30]

Although Huancavelica was unique in the colonial Americas as a source of mercury, it had much in common with silver mining towns beyond their role in sustaining Spain and serving as a catalyst for the rise of international trade networks. Among them was being known for unruliness, quick tempers, hard drinking, prostitution, theft and gambling. Its elite was known for being extravagant, indebted, and, like the overall population, male and young. This may help to explain why, in 1650, the city's 6,000 or so residents were estimated to consume between 3,300 and 6,600 gallons of wine annually. The vast majority of this was, however, consumed by the town's Spanish and Creole elite, who also enjoyed spirits, while the Indian population consumed copious quantities of chicha.[31] Initially, the gremio formed the town's economic elite, although by the 1650s they had been socially and economically eclipsed by the town's merchants, who also served as creditors. Contributing to the itinerant and volatile nature of the town were numerous *soldados*, or rootless fortune seekers. When they were not gambling, they conducted illegal mining operations, often chipping away at vital structural supports or using gunpowder to open new holes in the ground which could both damage mine workings and be a source of water penetration.[32]

29   Whitaker, 448; Contreras, *La ciudad del mercurio*, 43–44, 67, 75–76; Patiño Paúl Ortíz, 40; Ulloa, *Viaje a la América meridional*, 170; Pearce, "Huancavelica, 1700–1759," 694.

30   Contreras, *La ciudad del mercurio*, 42, 44–45; Patiño Paúl Ortíz, 38–39; Alejandro Reyes Flores, "Huancavelica, 'Alhaja de la Corona': 1740–1790," in *Ensayos en ciencias sociales*, Julio Mejía Navarrete, Ed. (Lima: Universidad Nacional Mayor de San Marcos, 2004), 61, 77–78 Povea Moreno, *Retrato de una decadncia*, 44, 48; Whitaker, 12.

31   Murua, 550; Patiño Paúl Ortíz, 43, 45, 47, 49, 51–52, 54–56, 62; Contreras, *La ciudad del mercurio*, 46–47, 52–53; Lohmann Villena, *Las minas de Huancavelica*, 116, 448; Whitaker, 12–13; Brown, "La crisis financiera," 366; Fisher, *Silver Mines and Silver Miners*, 11; Robins, *Mercury, Mining and Empire*, 32.

32   Mendoza y Luna, "Relación del estado del Gobierno," 34; Navarra y Rocaful, 170; *Descripción del virreinato del Perú*, 69; Contreras, *La ciudad del mercurio*, 58–59; Robins, *Mercury, Mining and Empire*, 44.

The eighteenth century was one of change in Huancavelica, as it was for much of the colony. A focus on more efficient administration, restricting contraband, greater technical expertise, and increased production for much of the period resulted in an overall, if fluctuating, increase in quicksilver yields. This period also saw four different systems of administration, beginning with the asiento/gremio system, and then followed by a contract with a sole producer, direct royal operation, and ultimately pallequeo. During this time, as with production, the population of the town fluctuated, however generally it increased. Although the town benefitted from public works, such as bridges and paved streets, Huancavelica remained a rowdy, restless and toxic town. It is to the nature and effects of such toxicity that we now turn.

CHAPTER 5

"They All Come to Die": Mining, Mishaps and Mercurialism[1]

It was in extraction and refining tasks that indigenous workers in Huancavelica most closely interacted with their environment. Initially, excavation from the Santa Bárbara mines was conducted through open pit mining. While this sounds better than being dispatched down a vent less shaft, it posed its own set of risks. As with subterranean mining, these workers ingested and inhaled considerable amounts of cinnabar dust and volatized mercury during their labors in what was some of the richest ore the hill had to offer. Less noxious but nevertheless constant, was dermal absorption of mercury by poorly clothed and shod workers covered in ore dust. Although open pit miners were not vulnerable to cave-ins, rock and landslides were a constant threat, most notably during the rainy months. Adding to worker's woes was frostbite from summer rains flooding the pit and winter freezes at 4,200 meters above sea level.[2]

Shaft Mining: "Crueler Than the Galleys [and]... Dungeons"[3]

By the turn of the 1600s, declining ore quality, and increasing problems with land and rockslides in the pit, led to the introduction of shaft mining. For miners, free and forced alike, things went from bad to worse. The entrance to the shaft was initially located at the bottom of the pit, and as water pours through a funnel, so too did it enter the shafts. While water had no problem entering the mine, people often did, due to rubble from rockslides which often blocked the entrance. Prior to the completion of the Bethlehem adit in 1642, workers accessed extraction points by descending a series of cactus-wood ladders connected to landings about twenty meters apart. Inside, the air was laden with carbon monoxide and thick with the smell of smoke, sulphur, sweat, excrement, and the sweet metallic scent of rich cinnabar. Compounding the dangers

---

1 Memorial del capitán Don Pedro Gutíerrez Calderón, 2.
2 Memorial del capitán Don Pedro Gutíerrez Calderón,1; Brown, "Worker's Health," 471; Robins, *Mercury, Mining and Empire*, 52.
3 Damian de Jeria, 2.

was arsenic gas which lurked in some reaches of the mine and could quickly kill a person. The warmth of the mine was a double-edged sword; for while it limited frostbite, it also facilitated the volatization of the mercury from the ore and associated dust. This, combined with an absence of ventilation, greatly increased the risks to which miners were exposed.[4]

The conditions were so horrific that the protector de los naturales in the town, Damían de Jeria, insisted that "those that enter [the mine] two or three times enter to die, from which it is clearly seen that shaft mining should not be permitted as it is a public slaughterhouse of so many Indians."[5]

The inescapable perils of shaft mining were clear to anyone who entered the shafts or saw those who worked them. In 1603, Friar Miguel Agia entered the San Jacinto and Mina Nueva mines to see the by now infamous conditions firsthand. His foray led him to urge the king to close the shafts and only engage in open pit mining. Similarly, Dr. Emetherio Ramírez de Arellano, a physician in Huancavelica writing in 1649, explained how "those who have sores on them... [get them as a] result of the excessive work of the loading and carrying [of the ore] more than twenty-four hundred feet...up, coming from an extreme heat to one of excessive cold, the combination of which results in the lung being compressed....giving rise to blood flow and illness...[and] some quickly died."[6] Such descriptions, while vivid, do not appear to be hyperbole, and impugn the sincerity of Viceroy García Sarmiento de Sotomayor, Conde de Salvatierra's, self-serving assertion in the mid-1650s that "in all the time of my administration I have not been given news...that there was any danger" to those engaged in mercury production.[7]

---

4   Memorial del capitán Don Pedro Gutíerrez Calderón,1; Emetherio Ramírez de Arellano, 1–2; Georgius Agricola, De Re Metalica. Herbert C. and Lou H. Hoover, Eds and Trans. (New York: Dover Publications, 1950), 428; Lohmann Villena, Las minas de Huancavelica, 182, 229; Patiño Paúl Ortíz, 349; Brown, "Worker's Health," 471, 474; Robins, Mercury, Mining and Empire, 52–53.

5   Damian de Jeria, 2, 5; José Sala Catala, "Vida y muerte en la mina de Huancavelica durante la primera mitad del siglo XVIII," in Asclepio 39 (1987), 195; Brown, "Worker's Health," 472–473, 475–477; Cook, Demographic Collapse, 205; Dobyns, 515; Enrique Tandeter, "Crisis in Upper Peru, 1800–1805," in Hispanic American Historical Review, Vol. VII1, No. 1 (1991), 66; Iden, Coercion and Market, 54; Robins, Mercury, Mining and Empire, 53, 56.

6   Emetherio Ramírez de Arellano, 1. See also Solórzano y Pereyra, Política Indiana, Vol. I, (Madrid: Matheo Sacristan, 1736), 131; Paulino Castañeda Delgado, "El tema de las minas en la ética colonial española," in La mineria hispana e iberoamericana, Vol. I (Leon, Spain: Catedra de San Isidoro, 1970), 345; Robins, Mercury, Mining and Empire, 135; Oñate, 141–142; Parés y Franqués, 84, 143, 144, 152; Brown, "Workers' Health," 472; Robins, Mercury, Mining and Empire, 135.

7   Sarmiento de Sotomayor, 245.

The risks of mining, and the injustice of using mitayos, were well known to King Philip III, (1578–1621) who in 1601 directed Viceroy Luis de Velasco to abolish the mita by attracting wage laborers and combining them with slave and convict labor.[8] As one colonial observer explained around this time, for those condemned to death, Huancavelica was "very well suited for [such] punishment."[9] Viceroy Velasco also recognized the perils of cinnabar mining, noting "the illnesses and deaths that the Indians…suffer without being able to avoid them… it is such that… when they excavate in the mines a dust is released…which settles in the chest….[and] causes a dry cough, a low fever, and in the end, death without repair."[10]

Although the exigencies of mercury production led Viceroy Velasco to abandon the effort to abolish the mita, he did reduce it to 1,600 people. At the urging of Damián de Jeria, he also ordered the shafts closed and allowed only open pit mining. The idea was that eventually this approach would result in a pit deep enough to reach the richest ore, while being somewhat less onerous on the Indians and with less "risk to the conscience of the one who orders them to work."[11] Good intentions and the production of quicksilver in Huancavelica did not go hand in hand, and as a result of Velasco's order, mercury production declined. Knowing that a viceroy's tenure was defined in no small way by minerals production, Velasco's successor, Gaspar de Zúñiga, Count of Monterrey, reopened the galleries in 1605. Shaft mining would prevail for the remainder of the colonial era, as production predictably trumped health.[12]

Small but important steps towards better conditions occurred with the arrival of Viceroy Juan de Mendoza, Marqués de Montesclaros, in Huancavelica. On August 2, 1608, he became the first viceroy to descend into the depths of the Santa Bárbara mines. Descending into the shafts from the open pit, the asphyxiating environment made quite an impression on him. According to the chronicler Francisco López Caravantes, during his visit "the first thing that [the viceroy] understood was that the opinion of those who said that these

---

8   Sala Catala, 194; Buechler, *The Mining Society of Potosí*, 43; Cole, 66; Ezquerra Abadia, 485; Fox, 64–65; Fisher, *Silver Mines and Silver Miners*, 212; Lohmann Villena, *Las minas de Huancavelica*, 185–187, 189; Patiño Paúl Ortíz, 212; Basadre, *El Conde de Lemos*, 123; Robins, *Mercury, Mining and Empire*, 68–69.

9   Memorial del capitán Don Pedro Gutíerrez Calderón, 2.

10   Carta de Virrey Velasco al Rey, Lima, May 5, 1600, AGI Lima 34, 6–7. See also Brown, "Worker's Health," 472.

11   Velasco, 112, 114; Jeria, 6.

12   Salinas y Córdoba, 328; Lizárraga, 211; Doña Nieves, 277; Lohmann Villena, *Las minas de Huancavelica*, 185–187, 189, 191, 197, 206; Brown, "Worker's Health," 472–473; Robins, *Mercury, Mining and Empire*, 53.

mines were a grave of Indians was correct."[13] As a result, the viceroy called for the excavation of ventilation tunnels to improve the air quality inside, nominally abolished night work, and prohibited mitayos from being detained for debts or unfulfilled quotas after their term was up. Seeking to improve the availability of medical care, he also increased the budget for the hospital.[14] Despite these well-meaning acts, the hospital offered little else but bleeding treatments and rest in lieu of mine or mill work, and improved airflow inside the mine would take decades.

Beyond mercury intoxication, workers inside the mine suffered silicosis, an incurable, and in acute cases, fatal, lung disease. Just as the limestone and sandstone contains cinnabar, it is likewise impregnated with silicone dioxide, or silica, which is released as dust when the rock is disturbed. When a person or animal breathes silica it damages, and ultimately scars, the lungs, even after exposure has stopped. This not only reduces the amount of oxygen that they can process and causes difficulty breathing, but triples their chances of contracting tuberculosis, especially among the undernourished. Initially, a person will suffer from a stubborn cough and fever before beginning to lose weight and have difficulty breathing. These symptoms may be accompanied by bluish skin and decreased kidney function, and in more severe cases, fibrosis, or excessive production of connective tissue.[15] As one writer in 1623 put it, after working in the mine, "those that are left live dying."[16]

The mortality of cinnabar mining and refining is underscored by the fact that from 1601 to 1604, of the 560 mitayos from the town of Huananhuancas, a dozen reportedly survived. Overall, it appears that even a short stint in Huancavelica was more deadly than work in Almadén, which tended to be for a longer term. This reflected Santa Bárbara's exceptional initial ore quality and lack of ventilation, as well as poor shaft construction, and extraction from structural supports which resulted in fatal cave-ins and land and rockslides.[17] Even after the improved access and ventilation which

13   Caravantes, Vol. IV, 175–176; Robins, Mercury, Mining and Empire, 55.

14   Montesinos, Vol. II, 189; Caravantes, Vol. IV, 176, 204; Robins, Mercury, Mining and Empire, 55.

15   Memorial del capitán Don Pedro Gutíerrez Calderón, 1–2; Emetherio Ramírez de Arellano, 1–2; Marvin Balaan and Daniel Banks, "Silicosis," in Environmental and Occupational Medicine. 3rd ed, William Rom, Ed. (New York: Lippincott-Raven Publishers, 1998), 438, 440–445; Brown, "Worker's Health," 475; Marvin Allison, "Paleopathology in Peru," in Natural History, Vol. VIII8, No. 2 (1979),82; Robins, Mercury, Mining and Empire, 55–56.

16   Memorial del capitán Don Pedro Gutíerrez Calderón, 1.

17   Marqués de Casa Concha, Relación del estado, 22; Memorial del capitán Don Pedro Gutíerrez Calderón, 1; Brown, "Worker's Health," 483, 491–495; Wiedner, 372; Cook, Demographic Collapse, 205; Robins, Mercury, Mining and Empire, 60.

resulted from the 1642 opening of the Bethlehem adit, and less toxicity as a result of declining ore quality after the loss of the main lode in 1645, about one-third of those in the mines would ultimately perish from the work, a figure which was likely considerably higher among refiners.[18]

Even in the 1840s in Almadén, where conditions were better, the consensus was that those working the mine face for six-hour shifts would exhibit signs of mercurialism within about ten days. Conditions did improve in Huancavelica in the eighteenth century, however, with the introduction of blasting, as opposed to crowbars, to dislodge ore. Although the use of gunpowder increased the dust and mercury vapor levels in the mine, improved ventilation through new adits which were completed in 1734 and 1760, helped flush them out.[19] With improved conditions in the 1720s, and less rich and lethal ore, came more even alquilas and the expansion of the Santa Bárbara village at the pithead.[20]

The dangers of mine work were not limited to poisoning from mercury, arsenic gas, silicosis and sharp temperature changes, but also included cave-ins and common accidents such as falling from ladders, collapsing platforms, and being hit by falling rocks. There was also the ever-present risk of *umpé*, or pockets of carbon monoxide or carbonic acid, which could kill a person before they could escape. Cave-ins generally resulted from two causes. One was the nature of cinnabar deposits in limestone and sandstone, which are found in pockets as opposed to veins. Consequently, the extraction of cinnabar leaves large voids in porous, and inherently unstable, rock, which can then give way to a cave-in or rockslide. Good mining practices can minimize these dangers, however as we have seen in Huancavelica the puentes and estribos were

---

18   Emetherio Ramírez de Arellano, 2–3; Brown, "Worker's Health," 495; Arana, 9; Salas Gue-
      vara, *Villa Rica de Oropesa*, 25.

19   Calancha, Vol. III, 879; Brown, "Worker's Health," 481–483, 484–486, 490, 495; Arana, 9; Sa-
      las Guevara, *Villa Rica de Oropesa*, 25; Rafael Dobado Gonzalez, "Salarios y condiciones de
      trabajo en las minas de Almaden, 1758–1839," in *La economía española al final del Antiguo
      Régimen. II. Manufacturas*, Pedro Tedde, Ed. (Madrid: Alianza Editorial/ Banco de España,
      1982), 361; Calancha, Vol. III, 879; Patiño Paúl Ortíz, 137, 144; Sala Catala, 196; Whitaker, *The
      Huancavelica Mercury Mine*, 18; Povea Moreno, *Retrato de una decadencia*, 260; Wiedner,
      372; Purser, 44; Robins, *Mercury, Mining and Empire*, 53–54, 56.

20   Calancha, Vol. III, 879; Brown, "Worker's Health," 481–483, 484–486, 490, 495; Arana, 9; Sa-
      las Guevara, *Villa Rica de Oropesa*, 25; Rafael Dobado Gonzalez, "Salarios y condiciones de
      trabajo en las minas de Almaden, 1758–1839," in *La economía española al final del Antiguo
      Régimen. II. Manufacturas*, Pedro Tedde, Ed. (Madrid: Alianza Editorial/ Banco de España,
      1982), 361; Calancha, Vol. III, 879; Patiño Paúl Ortíz, 137, 144; Sala Catala, 196; Whitaker, *The
      Huancavelica Mercury Mine*, 18; Povea Moreno, *Retrato de una decadencia*, 260; Wiedner,
      372; Purser, 44; Robins, *Mercury, Mining and Empire*, 53–54, 56.

routinely mined as, by the 1750s, they contained up to ten times as much cin-nabar as a gallery.[21]

Water intrusion from seasonal rains could also trigger a rockslide, such as the one on February 9, 1608. Almost five weeks later, on March 15, a cave-in from mining pillars injured the interim governor and took the life of two Span-iards, a veedor, and an overseer. Some accidents were especially deadly, such as that deep in the Santo Domingo de Cochapata mine, in which over 100 mi-tayos from Chumbivilcas perished in the early seventeenth century. Mining of critical structural supports also caused the landslide of October 6, 1616 in the "Sacadero" section of the mines, which ended the lives of at least a dozen people. In 1639 and 1640, several others perished in the Hojaldrado and Surtina sections, respectively, with the latter collapse reducing the mine's ventilation, access and production.[22] Similar accidents included those of 1681 in the San Jacinto zone, and one in 1759 which closed a ventilation shaft and the workings of the San Alejo section. These were followed in the same year by a collapse in the San Bruno sector, and yet another in the Las Animas, Santa Cruz, La Sole-dad and San José sections. Mining of supports also caused the 1760 rockslide in the Yerbabuena section, and that of 1761 in the San Antonio el Bajo mine. The worst cave-in, however, was that of September 25, 1786, which killed up-wards of 300 Indians, flooded the Bethlehem adit, and destroyed eight streets, twenty-eight plazas, 103 pit faces and 155 supports in the Brocal mine.[23]

---

21   Mendoza y Luna, "Carta del Virrey Marqués de Montes Claros," 89; Llano Zapata, 221; Na-varra y Rocaful, 163; N.A, "Memoria sobre la mina," 91–92; Arana, 9–10; Molina Martínez, *Antonio de Ulloa*, 77; Iden, "Tecnica y laboreo," 405; Patiño Paúl Ortíz, 71; NA, "Memoria sobre la mina de azogue de Huancavelica," 91–92; Patiño Paúl Ortíz, 135; Wise and Féraud, 18–19; Lohmann Villena, *Las minas de Huancavelica*, 218, 258–259, 326–327; Navarra y Ro-caful, 166; Ulloa, *Noticias Americanas*, 273–274; Arana, 38; Berry and Singewald, 22–23; Wise and Féraud, 19; Pearce, Adrian J. "Huancavelica 1700–1759," 677–678; Robins, *Mer-cury, Mining and Empire*, 56.

22   NA, "Memoria sobre la mina de azogue de Huancavelica," 91–92; Patiño Paúl Ortíz, 71, 135; Wise and Féraud, 18–19; Lohmann Villena, *Las minas de Huancavelica*, 218, 258–259, 326–327; Robins, *Mercury, Mining and Empire*, 56.

23   Llano Zapata, 221; Navarra y Rocaful, 163; N.A, "Memoria sobre la mina," 91–92; Arana, 9–10; Molina Martínez, *Antonio de Ulloa*, 77–78; Iden, "Tecnica y laboreo," 405; Fernández Alonso, 351; Arana, 31 Patiño Paúl Ortíz, 92; Bargallo, *La minería y metalurgía*, 341–342; Berry, *The Geology and Paleontology*, 24; Rivero y Ustariz, *Memoria Sobre El Rico Mineral De Azogue*, 49; Strauss, 562; Contreras and Díaz, 8; Arana, 10; Lang, 222–224; Arana, 11, 31; Berry and Singewald, 23; Wise and Féraud, 18–19; Fisher, *Silver Mines and Silver Miners*, 9; Patiño Paúl Ortíz, 71, 89, 91, 135; Salas Guevara, *Villa Rica de Oropesa*, 25; Whitaker, 66; Yates, et al., 19; Robins, *Mercury, Mining and Empire*, 56.

Overall, such collapses carried away at least 500 lives. It is, however, difficult to determine the number of fatalities, as overseers and other officials sought to cover them up to conceal illicit practices or not be held responsible for the deaths. In addition, while a mitayo's death would be noted as a result of his absence from muster, the death of a wage laborer would more likely go unnoticed as such workers, and often their production, were not registered.[24]

Venting Frustrations: The Bethlehem Adit

Apart from structural issues resulting from the nature of the rock, water intrusion and mining supports, the lack of ventilation and access to the mines contributed greatly to mercury poisoning. To remedy this, in 1609 workers began excavating the Our Lady of Bethlehem adit, which would not only provide better air circulation but also obviate the use of many ladders as carguiches ferried ore to the pithead. At first, however, it seemed like it may take an eternity to complete the project, as progress by pick-ax was excruciatingly slow. In 1635, after twenty-six years of effort, workers began to use gunpowder to blast their way forward. The results were remarkable, as the length of the adit doubled in the next four years. In April, 1642, to much fanfare, the new access was finally opened. Running over 457 meters inside the Santa Bárbara mount, the shaft stood about three meters high and almost as many wide.[25]

The completion of this project was a watershed in Huancavelica's mining history, although one contemporary's assertion that "they have today made a house of pleasure from that which caused horror for the lives that were lost there" is doubtful.[26] It did, however, provide markedly improved ventilation and ease of ingress and egress, thus reducing health and accident risks for all those entering the mine. Previously, a carguiche was expected to carry out one load of ore per day, but with the new adit they could carry forty, although mules were increasingly used as they carried six times what an Indian could.[27]

Better ventilation, combined with the loss of the main lode in 1645 and generally declining ore quality translated into better conditions for both workers

---

24   Wise and Féraud, 19; Patiño Paúl Ortíz, 242; Lohmann Villena, *Las minas de Huancavelica*, 276.

25   Montesinos, Vol. II, 189, 212; Lohmann Villena, *Las minas de* Huancavelica, 210, 332–333; Brown, "Worker's Health," 481–482; Robins, *Mercury, Mining and Empire*, 53–54.

26   Sala Catala, 199.

27   Emetherio Ramírez de Arellano, 2–3; Montesinos, 189; Lohmann Villena, *Las minas de Huancavelica*, 333–334.

and the environment. Governor Jerónimo de Sola proffered a remarkably rosy description of conditions, asserting that "No longer do they talk... of the fears they had before of becoming poisoned and losing their life or health; a disinterested person cannot deny that it used to be... common that there was not a pick man, generally, that could bear three or four years in the work without coughing up blood and getting poisoned; and now one sees them come and go [from the mine] as robust at the end of this time as the first day."[28] Dr. Ramírez de Arellano shared this view in 1649, noting the reduction of respiratory illnesses and accidents which followed the opening of the adit. He noted that "the mine is fresh, bathed with air which sweeps away the dust, and does not make the impression as before. And...the work is much less as they enter the mine walking and descend very little by ladders." The result was that most of the Indians who arrived at the hospital reportedly suffered from accidents and "common illnesses...such as fevers, flank pains, coughs and viruses," which miners and non-miners suffered.[29]

Over 100 years later, Governor Ulloa reinforced this view, averring that in the first few years of his administration only three or four Indians died in the mine. He wrote that in the mines

> the people who work... do not get sick, as is commonly believed: In the old days this harm was said to be more frequent and was attributed to two causes; one [is] the greater amount of mercury which the ore contained, another is the manner of breaking it in the mine with a pick, as the dust which was released was brought by breathing into their lungs, and caused them to sicken. Those that presently get sick are few, and these get it in the smelters when they load them, by entering before they have cooled; but as the ore is of poor quality, not even this is common.[30]

By 1765 the hospital administrator made the remarkable claim that in the past year no patients in the hospital had become ill from the mine.[31]

While there is probably an element of truth to these descriptions, these officials also sought to portray their administrations in a positive light. Given the subclinical, or unapparent, effects of mercury poisoning, the native's strong reluctance to enter the hospital, and tendency to relocate to recover, an absence of patients in the hospital does not necessarily indicate an absence of

28   Carrasco, 228; Molina Martínez, *Antonio de Ulloa*, 87.
29   Emetherio Ramírez de Arellano, 2–3.
30   Molina Martínez, *Antonio de Ulloa*, 88; Ulloa, *Noticias Americanas*, 281.
31   Povea Moreno, *Retrato de una decadencia*, 260–261.

illnesses. Entering a smoldering smelter emanating mercury vapor is going to have a harmful effect on the worker, even with poor quality ore. Huancavelica remained an unhealthy place, which even Ulloa acknowledged when he noted the prevalence of people coughing up blood.[32] In the early eighteen century, over sixty years after the Bethlehem adit was opened, the Council of the Indies held the view that anyone who worked for six months in the Santa Bárbara mine would suffer acute mercury poisoning.[33]

Helter Smelter

The refining of cinnabar presented additional hazards and technical challenges. Among them was the consistent provision of fuel for refiners. Wood such as queñua provided an excellent source, burning at a high temperature while still green. Supplies, however, were limited, and bringing them to town increased the costs of production. This sparked an interest in alternative fuel sources. *Taquia*, or llama excrement, was also used, but it had a low caloric value and burned quickly, thus requiring large volumes for smelting. The coal deposits of Pallalla, about forty kilometers from town, were another option, however extraction and transportation costs precluded their exploitation.[34]

It was in this context, that in 1570 Rodrigo de Torres de Navarra settled in Huancavelica and began to experiment with different fuels for smelters. In so doing, in 1572 he noticed that many Indians used ichu as a cooking fuel. His interest was piqued, and in subsequent experiments he demonstrated that it could be used to refine mercury. Despite the large quantity necessary for smelting, it had many advantages. These included the ubiquity of the plant on the altiplano, the ease with which it is gathered and transported, and the fact that it can be harvested every two years. Ichu-fueled smelting could also be done in the open, as the fire did not need much protection.[35]

---

32   Ulloa, *Noticias Americanas*, 210; Robins, *Mercury, Mining and Empire*, 137.

33   Brown, "Worker's Health," 483; Patiño Paúl Ortíz, 225, 258–259; Pearce, "Huancavelica 1700–1759," 681; Robins, *Mercury, Mining and Empire*, 70.

34   Cantos de Andrade, 304; Patiño Paúl Ortíz, 190–191; Arana, 102; Strauss, 562; Hawley, 8; Singewald, "The Huancavelica Mercury Deposits," 522; Carlos Contreras, "El reemplazo del beneficio de patio en la minería peruana, 1850–1913," in *Revista de Indias*, Vol. V9, No. 216 (1999), 406; Patiño Paúl Ortíz, 191; Robins, *Mercury, Mining and Empire*, 32–33.

35   Cantos de Andrade, 304; Reginald C. Enock, *The Andes and the Amazon: Life and Travel in Peru* (London: T. Fisher Unwin, 1907), 203; Patiño Paúl Ortíz, 191; Lohmann Villena, *Las minas de Huancavelica*, 52–55; Sánchez Gómez, "La ténica en la producción," 163–164.

Ichu also offered cost savings in mercury production. Prior to its adoption for smelting, it was less expensive for refiners to transport the ore to a smelter near a wood source than to transport the wood to the town. This not only entailed hard costs, but also provided ample opportunity for theft of the ore by transporters. The cost of harvesting and transporting ichu to Huancavelica was considerably less than transporting the ore to a fuel source, and as a result it became the common practice. Such was the demand for ichu that in 1589 Viceroy Fernando Torres de Portugal issued regulations to protect it, prohibiting the uprooting of the plant and its use as a pasture food. Despite the utility of ichu, like llama dung, it did not achieve a high enough temperature to release all of the mercury contained in the cinnabar.[36]

Nevertheless, it was ichu upon which refiners, and the kingdom, depended to render mercury from Andean cinnabar. The process was simple, highly toxic, inefficient, and has consequences in the city to this day. Although mercury was refined in Santa Bárbara village near the pithead, most smelters were concentrated around the town, and especially in the "*chunca horno*" or "ten ovens" section of today's San Cristóbal neighborhood where mitayos were assigned to live. When the ore-laden llama train arrived at its designated smelter, it was received by an overseer who would then release it to workers. They would then crush it to maximize the yield, which was determined by the amount of mercury produced per *cajón*, or sixty-eight kilogram load, of ore. To refine a load required a hefty four tons of ichu, or about forty llama loads.[37]

Initially, refiners would place a load of prepared ore into a ceramic vessel, cap it with a cone-shaped ceramic top, and seal it with clay, mud or a mixture of mud and ash. The load was then fired, releasing sulphur dioxide, and once the temperature of the ore had reached a temperature of 580 degrees Celsius, mercury was released. Some of the quicksilver condensed on the ceramic lid and was subsequently collected, along with mercury which had deposited among the ash and ore in the vessel. The system was made somewhat more efficient with the addition of a spout to the lid, which directed the mercury

36   Cantos de Andrade, 304; "Confirmación de Su Magestad," 191; Lohmann Villena, *Las minas de Huancavelica*, 54; Patiño Paúl Ortíz, 190–191; N.A., "Memoria sobre la mina de azogue de Huancavelica," 115; Strauss, 563; Robins, *Mercury, Mining and Empire*, 32–33.

37   Cantos de Andrade, 304; Patiño Paúl Ortíz, 39; Fuentes Bajo, 92; Eugenio Lanuza y Soltelo, *Viaje ilustrado a los reinos del Perú* (Lima: Pontífica Universidad Católica del Perú, 1998), 114; Arana, 26; Strauss, 564; Molina Martínez, *Antonio de Ulloa*, 84; Purser, 45; Whitaker, 14; Robins, *Mercury, Mining and Empire*, 57, 110, 126; Strauss, 563–564.

vapor through a ceramic condensation tube which passed through a water chamber before releasing the liquid mercury into a container.[38]

This system had two major limitations. One was the small size, and hence yield, of each smelter, and the other was efficiency in terms of the amount of mercury which escaped. Even with the addition of the spout and condensation tubes, probably at least thirty percent of the mercury was released into the atmosphere. Pedro Contreras, one of the first cinnabar prospectors, miners and regidors of Huancavelica, developed a system in 1596 in which up to forty vessels could be refining at the same time. This system, known as a *jabeca* smelter, consisted of a fire chamber which stood just over a meter high, and which had openings on the top which received the lower part of the refining vessels. Usually beginning in the early morning, two workers would manage three ovens, loading the vessels with ore and covering them with about five centimeters of humid, packed ash. Refiners then poked holes through the ash to facilitate the exit of the mercury vapor. The loaded vessels were then capped with spouted lids before being sealed with clay or a mud-ash mixture. The tops were then connected to condensation tubes, as was the practice in the older system. Subsequent to firing, which lasted about nine hours, mercury was collected from the tops of the containers, the ash below, and the ceramic or glass receptacles. Although this system was more efficient, it still used the same materials as the older system, and was no more efficient in terms of mercury loss during refining.[39]

Other innovations followed that of Contreras, such as constructing a large adobe chamber in which the cinnabar was fired, essentially replacing the ceramic vessel by a structure, although without condensation tubes. As the mercury volatized, it would rise and condense on the ceiling of the chamber, or fall back to the ground below. Although more efficient than the jabeca ovens, it was not efficient in human life, as workers were commonly sent in to collect the mercury from the roof, floor and ash before the chamber had cooled, exposing them to considerable amounts of mercury vapor.[40]

---

38   Purser, 45; Bidstrup, 5; Goldwater, 49; Robins, *Mercury, Mining and Empire*, 57. For a discussion of estimated mercury vapor concentrations in Huancavelica at different points of the colonial period, see Robins, *Mercury, Mining and Empire*, 113–126.

39   Cantos de Andrade, 304; Mendoza y Luna, "Relación del estado del Gobierno," 44; Montesinos, Vol. II, 128; N.A., "Memoria sobre la mina de azogue de Huancavelica,"130; Purser, 45–46; Lohmann Villena, *Las minas de Huancavelica*, 55–56, 117, 137; Brown, "Workers' Health," 479; Strauss, 563; Robins, *Mercury, Mining and Empire*, 57–58.

40   Brown, "Workers' Health," 479; Strauss, 563; Robins, *Mercury, Mining and Empire*, 58.

This system was soon replaced in 1633 by one which essentially combined the two before it. A prospector-physician named Lope de Saavedra Barba adapted the fire-chamber oven described above by adding condensation tubes which ran from near the top of the cone-shaped roof, through water, and to a receptacle. Although mercury still condensed on the roof, floor and walls, the use of the condensation tubes facilitated the more efficient collection of mercury. This was a major technological breakthrough, allowing more quicksilver to be produced with a given amount of fuel, and with less labor. Moreover, unlike its predecessors, the system could process larger quantities of ore at once and more effectively refined those of poorer quality. Saavedra Barba's method became known as a *busconil* smelter, so named because its inventor was a *buscón*, or prospector. Not only did the busconil become the primary means of refining mercury in Huancavelica, but, in a west to east technology transfer, in 1646 it was adopted in Almadén, where it was used to the 1920s. There it was known as a Bustamante smelter, as Juan Alonso de Bustamante had studied it in Huancavelica before introducing it in Spain with minor modifications. Although Saavedra Barba was denied recognition of his method in Almadén, he was awarded a two percent royalty of the quicksilver produced in Huancavelica.[41]

When using busconil smelters, the *oyaricos*, or smelter operators, which included both mitayos and wage earners, would prepare the ovens by loading them with ore and fuel, assemble and seal the various condensation tubes, close the smelter and ignite the charge. The roasting of the ore was done on a grate positioned about three meters above the heat source and about the same distance from the roof. The initial charge burned for about four to six hours before refiners reduced the air supply and allowed the coals to smolder and ultimately extinguish. Twelve to twenty-four hours after closing the air supply, the smelter had cooled and mitayos were sent into the chamber to gather any mercury inside on the ceiling, walls and ground, and clean out the ash. Women and children also assisted in cleaning the condensation tubes and making *bolas*, or balls of mercury-containing ash and the discharge from the tubes which were then re-smelted.[42]

---

41   Melchor Navarra y Rocaful, 175; Cantos de Andrade, 304; Llano Zapata, 218; Purser, 46–47; Brown, "Workers' Health," 479; Fuentes Bajo, 92; Matilla Tascón, Vol. II, 89–96; Patiño Paúl Ortíz, 184, 186–187; Carrasco, 168; Strauss, 563; Lohmann Villena, *Las minas de Huancavelica*, 139; Iden, "La minería en el marco del virreinato peruano: Invenciones, sistemas, técnicas y organización industrial," in *La minería hispana e iberoamericana*, Vol. I. N.A. (Leon, Spain: Catedra de San Isidoro, 1970), 652; Matilla Tascón, Vol. II, 89–96; Whitaker, 83; Fernández Alonso, 367; Robins, *Mercury, Mining and Empire*, 58.

42   NA, "Memoria sobre la mina de azogue," 129–130; Expediente promovido por don Nicolás de Sarabia y Mollinedo, 3; N.A., "Memoria sobre la mina de azogue de Huancavelica," 129;

In Almadén, the state doctor described that when workers came out from cleaning and reloading the smelters "the sweat, the dust and the smoke made them totally unrecognizable to their coworkers."[43] Almadén's laborers also viewed smelting as the worst task they could be assigned, with one convict explaining that "it was very dangerous for one's health because the smoke... causes... many... to lose reason and others are left poisoned... this work is the most dangerous for the health of the men."[44] Highlighting the risks they faced is a 1960's study of miners and refiners in nine California cinnabar mining and refining sites which found that all of the workers who were poisoned by mercury were engaged in refining.[45]

In Huancavelica, it was not just oyaricos who were poisoned, but also their families, as their wives and children made bolas and cleaned the tubes. So common was this that one contemporary noted, "the Indians are so adept in refining that boys of seven or eight years watch the smelters and know how to give them the right amount of heat, and tempering the tops in time... with... water."[46] In an effort to minimize these risks, smelter groupings were to be separated to limit the amount of smoke to which Indians were exposed. In each smelter site, refiners were to have at least three retorts: one which was ready to load, another smelting, and the third cooling for a theoretical twelve or twenty-four hours, depending on if ichu or wood had, respectively, been used as a fuel.[47]

Part of the risk associated with the busconil smelters derived from their inefficiency. The smelter structure was not airtight, and mercury vapor escaped through cracks or fissures in the smelter as well as through the condensation

---

    Llano Zapata, 218; Gastelumendi, 55; Purser, 46–47; Brown, "Workers' Health," 479; Fuentes Bajo, 92; Matilla Tascón, Vol. II, 89–96; Patiño Paúl Ortíz, 184, 186–187; Carrasco, 168; Strauss, 563; Lohmann Villena, *Las minas de Huancavelica*, 139; Iden, "La minería," 652; Whitaker, 83; Fernández Alonso, 367; Strauss, 564; Povea Moreno, *Retrato de una decadencia*, 232; Brown, "Worker's Health," 479–480; Robins, *Mercury, Mining and Empire*, 57–60, 72, 136, 181.

43   Parés y Franqués, 93.

44   Menéndez Navarro, 81.

45   Irma West and James Lim, "Mercury Poisoning Among Workers in California's Mercury Mills," in *Journal of Occupational Medicine*. Vol. Io, No. 12 (December, 1968): 697–698.

46   Llano Zapata, 218; N.A., "Memoria sobre la mina de azogue de Huancavelica," 130; Strauss, 564; Brown, Worker's Health, 480; Robins, *Mercury, Mining and Empire*, 72.

47   King Charles II, "Confirmación de Su Magestad del Asiento que hizo el Excelentísimo Señor Duque de la Palata con los mineros de Guancavelica sobre la labor, y beneficio de la mina de azogue con las condiciones y calidades que se refieren." Madrid, June 10, 1685, in Minas e indios del Perú, siglos XVI–XVIII, Nadia Carnero Albarrán, Ed. (Lima: Universidad de San Marcos, 1981), 190; Robins, *Mercury, Mining and Empire*, 38.

tubes and the various connection points.[48] Huancavelica's tempestuous climate also played a role in the release of mercury vapor to the environment, and the poisoning of workers, underscoring the poor construction of the smelters. A nineteenth century description by Mariano Rivero y Ustariz described how wind could blow into the distillation tubes and out of the smelter door, blasting the workers with mercury vapor and ash. Sometimes the opposite happened, where wind would blow in through the door, overheating and rupturing the distillation tubes.[49]

This last point highlights that it was not only the smelters which were responsible for releases of mercury vapor into the atmosphere, so too were those who operated them. During the firing phase, refiners working at night would often sleep instead of keeping them loaded with the appropriate amount of fuel. To conceal this, they would overload the smelter with ichu so that their overseer would not see a large amount of it next to them; a sure sign they had been remiss in their task. This practice led to the expansion of the condensation tubes, the rupturing of the delicate seals along the joints, and considerable losses of mercury vapor.[50] Other sources of mercury vapor escape included improperly sealed inspection holes, through which refiners inserted a muddy stick which, when removed, indicated that firing was complete if it was not speckled with mercury. The risks to smelter operators were compounded, however, by the common practice of *endiabladas*, or opening the smelters to collect the mercury inside before they had sufficiently cooled. Although this could give the worker a debilitating dose of quicksilver vapor in an instant, it was nevertheless a common practice as overseers sought to meet production quotas.[51] Among those disproportionately affected appears to be Indians from Chumbivilcas. Not only were they the only district still sending men instead of money for the mita in the late 1700s, but they were dedicated to smelting.[52]

The design, construction and operation of the smelters resulted in an almost inescapable exposure to elemental mercury and its vapor not only by those who operated them, but all residents of the town. Just how much mercury

48   Jiménez, 206; Berry and Singewald, 27; Singewald, *The Huancavelica Mecury Deposits*, 522.

49   N.A., "Memoria sobre la mina de azogue de Huancavelica," 134; Robins, *Mercury, Mining and Empire*, 59. See also Crosnier, 55–56.

50   N.A., "Memoria sobre la mina de azogue de Huancavelica," 129–132; Fuentes Bajo, 93–94; Strauss, 564; Robins, *Mercury, Mining and Empire*, 59.

51   "Confirmación de Su Magestad," 191; Cantos de Andrade, 304; N.A., "Memoria sobre la mina de azogue de Huancavelica," 131–133, 136; Strauss, 564; Fuentes Bajo, 94; Brown, "Workers' Health," 479–480; Robins, *Mercury, Mining and Empire*, 38–59.

52   Povea Moreno, *Retrato de una decadencia*, 210, 215.

was lost to the air in the refining process is, however, difficult to determine, with estimates running from ten percent all the way to sixty percent.[53] It appears that losses were over thirty percent given that when refining resumed in Huancavelica in the early twentieth century utilizing improved smelters similar to those in New Almaden, California, losses were estimated at around fifteen to twenty percent. There is also a distinction between mercury vapor which is lost in refining and mercury which is left unrefined in tailings. Taking this latter consideration into account, and referring to the more efficient furnaces in Idrija, Slovenia, Strauss argues that eighty percent of the mercury was never captured, either as it was unrefined due to poor fuels or because it escaped as vapor.[54]

The amount of quicksilver produced in a firing depended on the richness of the ore, the integrity of the smelter, the heat of the fuel and the expertise of the refiner. The richest ores could contain thirty-percent mercury and would produce twenty kilograms of mercury per cajón, although five percent mercury, or a three and one half kilogram yield, or less, was more common. Generally declining ore quality meant that, by the late nineteenth century, most ore averaged only about one percent mercury.[55]

The refining of cinnabar released not only elemental mercury vapor, but also considerable amounts of sulphur along with the smoke from the fuel. Depending on the weather, thermal inversions could trap them over the town, and the toxic, malodorous air in Huancavelica was too much for some to handle. In June, 1735, the Franciscan Eugenio Lanuza y Sotelo visited Huancavelica in the company of other friars. After being ceremoniously received by Governor Sola y Fuente, the group decided to

> not go up to see the mine...we did not dare to because the mercury vapors... harmed us considerably, as there are so many smelters, which continually exhale fatal effluents such that one chokes and loses the ability

53 N.A., "Memoria sobre la mina de azogue de Huancavelica," 136; Gastelumendi, 56; Strauss, 565; Umlauff, 47; Carlos P. Jiménez, "Estadistica Minera en 1917," in *Boletín del cuerpo de ingenieros de minas del Peru*. No 95 (Lima: Imprenta Americana, 1919), 203; Arana, 35–36; Fuentes Bajo, 93; Menéndez Navarro, 127.

54 Berry and Singewlad, 27; Singewald, 522; Carlos Jímenez, 203; Strauss, 565.

55 Arana, 3, 26; Strauss, 562, 564; Berry and Singewald, 42, 46; Umlauff, 43–44; Wise and Féraud, 22; Molina Martínez, *Antonio de Ulloa*, 84; Purser, 45; Whitaker, 14; Tamayo, 40; Hawley, 8; William Hadley, "Report on the quicksilver mines of Huancavelica," in *Boletín del Ministerio de Fomento*, Vol. II, No. 1 (Lima, 1904), 44; Berry and Singewald, 46; Robins, *Mercury, Mining and Empire*, 57, 110.

to walk. For this reason... we continued our march [to Cuzco] due to the notable effect these vapors had on our Prelate.[56]

Once mercury had been refined, every two weeks it was weighed and inspected by an overseer at the smelters before being sent to crown's warehouse on the west side of the town's plaza. There, losses continued, as it was again weighed and stored in glazed but poorly fired ceramic vessels until it was poured into containers for transport to silver mining centers. Those who worked in the warehouse were constantly exposed to toxic vapors, resulting from spills during the various transfer processes, seepage through the ceramic containers, and breakage. Mercury is especially prone to splashing and forming very small, and sometimes invisible, droplets when being poured. In so doing, the total surface area can increase over thirty times, increasing the exposure risk of anyone present. Such spillage may have been the source of the almost one and one-half metric tons of mercury encountered during an excavation of the plaza in 1915, although it could also simply have been native mercury.[57]

A Slippery Cargo

The reach of Huancavelica's mercury, and its toxic effects, was far and wide. Not only did it supply nearby mines such as Cerro de Pasco, Lucanas, Castro-virreyna and Hualgayoc, but also those of Cailloma, Chucuito, Carangas, Oruro and Potosí. Beyond the Andes, Huancavelica's quicksilver succored mills in New Spain and Guatemala when supplies there ran short.[58] Some mercury was

---

56 Lanuza y Soltelo, 114.

57 Liñan y Cisneros, 313; Cantos de Andrade, 307–308; Patiño Paúl Ortíz, 35, 100; Carrasco, 127; Contreras, *La ciudad del mercurio*, 51; Lohmann villena, *Las minas de Huancavelica*, 415–416; Bidstrup, 6, 37–42; Robins, *Mercury, Mining and Empire*, 60, 182.

58 Manso de Velasco, 165–172, 311–312; Molina Martínez, *Antonio de Ulloa*, 110; Arana, 81; Lohmann Villena, *Las minas de* Huancavelica, 334; Cobb, *Potosí and Huancavelica*, 49; Whitaker, 52, 54; Guillermo Mira Delli-Zotti, "El Real Banco de San Carlos de Potosí y la minería Altoperuana colonial, 1779–1825," in *La savia del imperio: tres estudios de economia colonia*, Julio Sánchez Gómez, Guillermo Mira Delli-Zotti and Rafael Dobado, Eds. (Salamanca, Ediciones Universidad Salamanca, 1997), 332; Rafael Dobado González, "Las minas de Almadén, el monopolio del azogue y la producción de plata en Nueva España en el siglo XVIII," in *La savia del imperio: tres estudios de economía colonial,* Julio Sánchez Gómez, Guillermo Mira Delli-Zotti and Rafael Dobado, Eds. (Salamanca, Ediciones Universidad Salamanca, 1997), 470; Matilla Tascón, Vol. II, 391–392; María Dolores Fuentes Bajo, "El azogue en las postrimerías del Perú colonial," in *Revista de Indias*. Vol. XLVI (January-June, 1986),

released to the environment through losses during transport, although it was primarily volatized in silver refining towns. After 1575, when the crown ceased transporting mercury directly, the delivery of quicksilver to silver mining centers was contracted out. Prior to this, however, mercury stored in the royal warehouse was prepared for transport by placing it in clay vessels which could hold either a half-quart or quart of mercury, or about five to ten kilograms. The problem with these containers was that they were prone to leaking and breakage. To avoid such losses, transporters began to use quart-sized sheepskin bags, usually produced in Chile. Although they were more flexible and utilized for most of the colonial era, leakage or rupture were perennial problems and poisoned the llamas and mules who carried them.[59]

The initial route of transport to Upper Peru was via Cuzco by llama train, although in the 1580s contractors began to send it to Pisco. There it was loaded onto ships which ferried it down the coast before bringing it to the highlands via Arequipa. In the mid-1590s, the shipping route was changed somewhat, with the quicksilver being loaded at Chincha, near Pisco, for shipment to Arica, and thence to the highlands. On their way down to the coast, transporters stopped at San Gerónimo, about forty-four kilometers away. Here, the cargo was placed on mules for the final descent to the hot, dry coast. On January 24,1666 this royal warehouse caught fire, vaporizing around forty metric tons of mercury, most of which ultimately settled in the region.[60]

From Arica, transporters working under a separate contract loaded the mercury onto mule trains which carried it about seventy-two kilometers upland before it was transferred to llamas. Given their stamina and the load they can carry, after around 1610 mule trains were generally utilized for the trip to highland mining towns such as Oruro and Potosí. Following the truncation of the Viceroyalty of Peru and the establishment of that of Rio de la Plata in 1776, Huancavelica's production was dedicated to Peru's mills.[61]

102; Pearce, "Huancavelica 1700–1759," 698; Molina Martínez, *Antonio de Ulloa*, 96; Lohmann Villena, *Las minas de* Huancavelica, 165; Arana, 81; Robins, *Mercury, Mining and Empire*, 10.

59   Cobb, *Potosí y Huancavelica*, 113–114; Iden, "Supply and Transportation," 40, 42; Roel Pineda, 103–104; Lohmann Villena, *Las minas de* Huancavelica, 109, 161; Patiño Paúl Ortíz, 35, 160; Basadre, *El Conde de Lemos y su tiempo*, 158; Robins, *Mercury, Mining and Empire*, 70. For a list and discussion of the transport asientos, see Zavala, Vol. I, 225–228.

60   Audiencia de Lima, "Audiencia de Lima al Señor Virrey Conde de Lemos," 204; Basadre, *El Conde de Lemos y su tiempo*, 158.

61   Carta de su presidente, el licenciado Alonso Maldonado de Torres, asistente en Potosí, a esta Real Audiencia: Recomienda el favorecer la conservación de la mita sin desmedro ninguno. Potosí, December 2, 1606. ABNB, ALP Minas, 123/2, 1; Mendoza y Luna, "Relación

Refining Estimates

Colonial officials in Huancavelica kept account of the mercury which gremio members, their lessees and their creditors brought to the royal warehouse to be exchanged for silver. From 1570 to 1820, estimates of mercury production in Huancavelica vary from 48,000 to 62,000 metric tons.[62] This figure does not, however, include contraband, which diverted tons of the liquid metal from the royal monopoly. Such diversion is inherently difficult to determine as it is undocumented. What we do know is that Huancavelica was notoriously corrupt, the vortex of a well-entrenched graft network stretching from civil officials in the mines to viceroys in Lima. Beyond their own involvement, and financial interest, in the illicit production and sale of mercury, officials were hesitant to carry out a sustained crackdown out of concern of the gremio's collective response, which could involve a work stoppage.[63]

del estado del Gobierno," 45–47; Montesinos, Vol. II, 94; Murua, 550, Acosta, 245; Cantos de Andrade, 306; José de la Riva Aguero, "Descripción anónima del Perú y de Lima a principios del siglo XVII," in *Revista Historica*, Vol. II1, (Lima, 1954), 35; Bargallo, *La amalgamación*, 344; Patiño Paúl Ortíz, 160; Cobb, *Potosí y Huancavelica*, 112–114, 117–118; Iden, "Supply and Transportation," 40, 42; Alvaro Jara, *Tres ensayos sobre economía minera hispanoamericana* (Santiago: Universidad de Chile, 1966), 73; Expediente sobre la postura para la conducción de azogue de esta Caja de Potosí a las demas del Virreinato. Potosí, 1781. CNMAH, CGI/M-64/26, 1; Patiño Paúl Ortíz, 89; Basadre, *El Conde de Lemos y su tiempo*, 158; Robins, *Mercury, Mining and Empire*, 71.

62   Caravantes, Vol. IV, 224; Haenke, 266; Rivero y Ustariz, 20, 45–47; Arana, 82–83; Loveday, 82; Jiménez, 203; Patiño Paúl Ortíz, 163–164; Brown,"La crisis financiera," 352, 366, 375, Whitaker, *The Huancavelica Mercury Mine*, 6–7, 104; Roel Pineda, 105; Berry and Singewald, 16; José María Blanco, *Diario de viaje del Presidente Orbegoso al sur del Perú*. Vol. I, Felix Denegri Luna, Ed. (Lima: PUCP-IRA 1974), 37; Robins, *Mercury, Mining and Empire*, 110. See also Robert Yates, Dean F. Kent, and Concha J. Fernández, *Geology of theHuancavelica Quicksilver District, Peru* (Washington, D.C.: U.S. G.P.O, 1951), 1.

63   Montesinos, 74, 212; Wiedner, 369; Fernández de Castro y Andrade, Advertencias que hace el Conde de Lemos, 255; Sarmiento de Sotomayor, 245 Whitaker, *The Huancavelica Mercury Mine*, 41; Octavio Puche, "Influencia de la legislación minera, del laboreo, así como del desarrollo técnico y ecnonómico, en el estado y producción de las minas de Huancavelica, durante sus primeros tiempos," in *Minería y metalurgia: Intercambio tecnológico y cultural entre América y Europa durante el período colonial español*. Manuel Castillo Martos, Ed. (Seville: Muñoz Moya y Montraveta, editors, 1994), 469; Cobb, *Potosí and Huancavelica*, 25;

Lohmann Villena, *Las minas de Huancavelica*, 82,174, 254, 267, 284, 286, 470; Lohmann Villena, *Las minas de Huancavelica*, 256, 267, 452; Molina Martínez, *Antonio de Ulloa*, 74; Whitaker, 41, 43; Patiño Paúl Ortíz, 161; Abrines, 107; Fernández Alonso, 347; Pearce, "Huancavelica 1700–1759," 677; Robins, *Mercury, Mining and Empire*, 111–112.

With robust demand in nearby silver mining centers, such as Castrovirrey-na, Cerro de Pasco and Hualgayoc, Indian miners in Huancavelica were well-placed to extract illicit cinnabar and sell it as ore or refine it. Gremio members neither forgot nor forgave Viceroy Toledo's expropriation of their claims, and had little incentive to hand over their entire production to local officials, who in turn would take their cut. Such was the corruption and disarray in the trea-sury offices that in 1668 Viceroy Lemos described their accounting as "almost incomprehensible."[64] Beyond excavation, transporting ore to smelters, refin-ing, and depositing mercury in the royal warehouse, transport of mercury to silver refining sites offered additional opportunities for diverting mercury to the black market.[65]

Complicating the situation was the scarcity of specie in the town, with the result that mercury was often used as a currency by miners and merchants alike. Although periodic crackdowns on the illegal traffic in mercury had some, albeit short-lived, effects, in the words of Governor Juan Luis López such illicit commerce was "almost a necessary evil" to ensure both mercury and silver pro-duction.[66] The proximity of silver mines in the region were frequent destina-tions of Huancavelica's illicit production, resulting in an estimated twenty-five percent of the silver produced in these centers being unregistered. According to one calculation, the crown lost 292 and one half pesos for every forty-five kilograms of mercury diverted from the treasury, a sum that reflects lost profits on the mercury as well as the lost tax on the silver produced with it. Gener-ally, contraband rates varied inversely with production levels, and directly with delayed remittances from Lima, increasing in times of diminished production and tardy silver shipments.[67] Another illegal, albeit tolerated, practice was the

---

64   Cueva, 186; Fernández de Cabrera, 90; N.A., *Memoria sobre la mina de* azogue, 117; Manso
      de Velasco, 159, 161; Armendaris, 162; Navarra y Rocaful, 175; Molina Martínez, *Antonio
      de Ulloa*, 109; Brown, "La crisis financiera peruana," 364–365, 375; Patiño Paúl Ortíz, 114;
      Lohmann Villena, *Las minas de Huancavelica*, 276; Bakewell, "Registered Silver Produc-
      tion," 84; Puche, 469–470; Álvarez de Toledo y Leyva, 172; Castro y Andrade, 255; Robins,
      Mercury, Mining and Empire, 112.

65   Molina Martínez, *Antonio de Ulloa*, 108.

66   Carta de Pedro Camargo, 2; Brown, "La crisis financiera," 366, 375, 379–380; Patiño Paúl
      Ortíz, 74.

67   Mendoza y Luna, "Carta del Virrey Marqués de Montes Claros a S.M. informando extensa-
      mente sobre las minas de Guancavelica," 96; Audiencia de Lima. "Relacion que la Real Au-
      diencia de Lima hace," 295; Manso de Velasco, 159; Álvarez de Toledo y Leyva, 144; Navarra
      y Rocaful, 164–165, 175; Fernández de Cabrera, 90; Armendaris, 162, 362; Cueva, 186; Manso
      de Velasco, 159; Rivero y Ustariz, *Memoria Sobre El Rico Mineral De Azogue*, 20; Fisher, *Sil-
      ver Mines and Silver Miners*, 6, 110; Iden, "Silver Production in the Viceroyalty of Peru," 287;

purchase of mercury by merchants from producers at a discounted price of forty pesos, and its subsequent resale to the government at the official rate of fifty-eight pesos when funds had arrived. While technically this was a form of contraband, the practice was accepted by civil authorities as at least a portion of the mercury eventually ended up in the royal warehouse.[68]

Contemporaries and later scholars have surmised that between one-tenth and two-thirds of the mercury produced in Huancavelica was unregistered. Given this range, and the context of rife corruption, a conservative estimate of unrecorded mercury production is twenty-five percent of the total registered. Added to the official production records, this yields a total production of about 68,200 metric tons between 1570 and 1820.[69] In 1810 the Spanish Cortés abolished the mercury monopoly and allowed the unrestricted sale of quicksilver. With this came the end of centralized recordkeeping of production until the twentieth century. Several studies, however, have estimated production during this time. Combined with twentieth century production records, approximately 5,913 metric tons have been produced in Huancavelica during the national period, yielding a grand total of about 72,125 metric tons.[70] Presently in the

Gaston Arduz Eguía, *Ensayos sobre la historia de la minería altoperuana* (Madrid: Paraninfo, 1985), 71; Fernández Alonso, 357–358; Bakewell, "Registered Silver Production," 81; Contreras, *La ciudad del mercurio*, 108; Brown, "La crisis financiera," 364, 370, 375; Patiño Paúl Ortíz, 114–115; Molina Martínez, *Antonio de Ulloa*, 109; Fernández Alonso, 349, 362; Pearce, "Huancavelica," 675; Robins, *Mercury, Mining and Empire*, 112.

68   Manso de Velasco, 159; Brown, "La crisis financiera," 366, 370; Lohmann Villena, *Las minas de Huancavelica*, 470; Pearce, "Huancavelica," 675–676.

69   Arana, 82; Brown, "La crisis financiera," 352, 375, Whitaker,104; Roel Pineda, 105; Cross, 408; Patiño Paúl Ortíz, 163–164; Fernández Alonso, 349; Robins, *Mercury, Mining and Empire*, 110.

70   Whitaker, 81; Wise and Féraud, 23; Arana, 82; Dueñas, 156; Berery and Singewlad, 16; Government of Perú, *Inventario Y Evaluación De Los Recursos Naturales De La Zona Altoandina Del Perú: Reconocimiento, Departamento De Huancavelica*. Vol. I. (Lima: Oficina Nacional de Evaluación de Recursos Nacionales, 1984), 181; Iden, Government of Peru, *Centros Poblados: Sexto Censo Nacional De Población, Primer Censo Nacional De Vivienda, 2 De Julio De 1961*. Vol. II (Lima: Dirección Nacional de Estadísticay Censos, 1966), 260; Wise and Féraud, 23. For differing calculations, see Loveday, 83, Whitaker, 83, Crosnier, 57; Strauss, 562; Yates, et al., 21; Jiménez, 17, 21; Wise and Féraud, 15; Umlauff, 41; Mariano de Rivero y Ustariz, *Memoria Sobre El Rico Mineral*, 20.45–47; Carlos Contreras and Ali Díaz, *Los intentos de reflotamiento de la mina de azogue de Huancavelica en el siglo XIX* (Lima: NP, 2007), 11–12; See also NA. "Historia de la mina de Huancavelica," in *Mercurio Peruano* (January 30, 1791), 67; Clements Markham, *Cuzco: A journey to the Ancient Capital of Peru; with an Account of the History, Language, Literature, and Antiquities of the Incas. And Lima: A Visit to the Capital and Provinces of Modern Peru* (London: Chapman and Hall, 1856), 54; Enock, 131;

region of Huancavelica there is only artisanal mercury production of mercury, producing approximately one-half metric ton a year.[71]

Toxic Burdens

The toxicity of mercury largely depends on its species, dose, length of exposure and how it is ingested. Individual responses vary, as some people are genetically more sensitive than others, and those who are malnourished or otherwise in poor health are more susceptible to its effects. Age and gender are also factors, and women and children are especially vulnerable. Exposure to mercury can affect women's fertility and menstrual cycles and result in stillbirths or mental and physical birth defects, even where there is no obvious sign of mercury intoxication.[72] Mercury's effects on children are exacerbated as a result of

---

Helena Meyer and Alethea Mitchell, "Mercury," in *Minerals Yearbook 1950* (Washington, DC: Bureau of Mines, 1953), 785; Helena Meyer and Gertrude Greenspoon, "Mercury," in *Minerals Yearbook: Metals and Minerals (except fuels) 1953* (Washington, DC: Bureau of Mines, 1956), 784; J.W. Pennington and Gertrude N. Greenspoon, "Mercury," in *Minerals Yearbook: Metals and Minerals (except fuels), 1956*. Vol. I. (Washington, D.C.: Bureau of Mines, 1958), 826; Iden, "Mercury," in *Minerals Yearbook: Metals and Minerals (except fuels), 1958*. Vol I (Washington, D.C.: Bureau of Mines, 1959), 761; John Burgess, Sumner Anderson and R. Lester, Jr., "Peru," in *Minerals Yearbook Area Reports: International 1963*. Vol. IV. (Washington, DC: Bureau of Mines, 1964), 306; Lester Brown, Jr, "Peru," in *Minerals Yearbook Area Reports: International, 1967*. Vol. IV (Washington, DC: Bureau of Mines, 1969), 604; Frank Noe, "Peru," in *Minerals Yearbook Area Reports: International 1969*. Vol. IV (Washington, DC: Bureau of Mines, 1969), 575; F.W. Wessel, "Peru," in *Minerals Yearbook Area Reports: International 1972*. Vol. III (Washington, DC: Bureau of Mines, 1972), 640; Iden, "Peru," in *Minerals Yearbook Area Reports: International 1974*. Vol. III (Washington, DC: Bureau of Mines, 1974), 718; Orlando Martino, *Minerals Yearbook Area Reports: International 1976*. Vol. III (Washington, DC, Bureau of Mines, 1976), 824.

71   Wise and Féraud, 23.

72   "Health Effects," 67, 161, 221, 278; USEPA, *Mercury Study*, USEPA, *Mercury Study*, 2–6; Andrew Rowland, et al.,"The effect of occupational exposure to mercury vapour on the fertility of female dental assistants," in *Occupational and Environmental Medicine*. Vol. V1, No. 1 (January 1994), 28–34; Keith Yeates and Mary Ellen Mortensen, "Acute and Chronic Neuropsychological Consequences of Mercury Vapor Poisoning in Two Early Adolescents," in *Journal of Clinical and Experimental Neuropsychology*. Vol. 16, No.2 (1994), 209–210, 218; USEPA, *Mercury Study*, 0–3, 2–6; Kirsten Alcser, et al.,"Occupational Mercury Exposure and Male Reproductive Health," in *American Journal of Industrial Medicine*, Vol. I5, No. 5 (1989): 517; S.M. Barlow, et al., "Reproductive hazards at work," in *Hunter's Diseases of Occupations*. 8th ed. P.A.B. Raffle et al., Eds. (London: E. Arnold, 1994), 729; "Health effects," 188, 219, 223, D'Itri, 26; Schutte, 553; Bakir, F., et al. "Methylmercury Poisoning in Iraq," in

greater exposure from mouthing activity, higher concentrations due to smaller body size, and because their brain and central nervous system are developing. Irrespective of the characteristics of the person who is intoxicated, mercury in its various forms affects the central nervous system and also accretes in the kidneys. For example, autopsies of twentieth-century cinnabar miners in Idrija, Slovenia, found elevated mercury concentrations in their brains, kidney cortex, thyroids and pituitary glands.[73]

Elemental mercury, such as that produced in Huancavelica, is liquid at ambient temperatures. If applied to the skin, about two percent will enter the body, and if it is consumed orally, about ten percent will enter through the gastrointestinal tract before the remainder is excreted. Much more toxic, however, is breathing elemental mercury vapor, as did the oyaricos and other inhabitants of colonial Huancavelica. When inhaled, approximately eighty-five percent of mercury vapor will remain in the body and settle in the brain, kidneys and other organs. Once inside the body, its half-life is approximately two months as it is slowly eliminated through urination, feces and perspiration. It is also excreted in mother's milk, and a baby who consumes contaminated milk has probably already been poisoned in the womb.[74]

Elemental mercury can also combine with minerals and organic matter to form compounds. Organic compounds result when elemental mercury binds with carbon atoms, while inorganic compounds occur when mercury combines with minerals such as salts, sulfates, sulfides, nitrates and chlorides. The action of sunlight on elemental mercury vapor also causes it to transform into divalent mercury, an inorganic form. This occurs when elemental mercury volatizes and enters the atmosphere, where it will often bind to particulate matter before depositing on the earth's soils or waterways.[75]

Science, Vol. I81, No. 4096 (July 20, 1973), 234, 238–239; Bidstrup, 8, 34; Waldron, 103; Hugh Evans, "Mercury," 998; Rowland, 28; Robins, Mercury, Mining and Empire, 106, 136.

73   Schutte, 553; D'Itri, 135; "Health Effects," 327; Parés y Franqués, 104; Yeates and Mortensen, 209; Evans, "Mercury," 1001; I.M. Falnoga, Tusek-Znidaric, et al., "Mercury, Selenium, and Cadmium in Human Autopsy Samples from Idrija Residents and Mercury Mine Workers," in Environmental Research, Vol. VIII4, Section A, (2000), 211, 213; Robins, Mercury, Mining and Empire, 104–105, 108, 136.

74   Hugh Evans, "Mercury," 1000; Waldron, 91, 102, 105; Schutte, 552; Waldron, 103; Bidstrup, 8, 43, 60, "Health Effects," 31, 33, 47–49, 54–55, 58, 161–162, 163,185–186, 264–265, 268, 272; USEPA, Mercury Study, 0–3; D'Itri, 26; Robins, Mercury, Mining and Empire, 104, 106.

75   D'Itri, 105, 118; Goldwater, Mercury, 127–128; David O. Marsh, "Organic mercury: methyl-mercury compounds," in Handbook of Clinical Neurology. Vol. III6, P. Vinken and G. Bruyn, Eds. (New York: Elsevier, 1979), 73; Waldron, 102; USEPA, Mercury Study, 0–2, 3–13; USEPA, Mercury Study, 2–1; Robins, Mercury, Mining and Empire, 103–104.

Inside the body, organic and inorganic compounds behave differently. Whereas inorganic compounds bind with both plasma and red blood cells and target the kidneys, organic compounds bind primarily with red blood cells and more easily penetrate the brain and central nervous system. The effects of organic mercury compounds are exacerbated by the fact that when they are orally ingested, unlike elemental mercury, at least ninety percent will be absorbed by the gastrointestinal tract and ultimately enter the bloodstream and brain. For this reason, organic mercury compounds are especially toxic, and damaging to the central nervous system. Severe motor problems, tremors and slurred speech often result, although unlike elemental mercury poisoning, there is less prevalence of hyper salivation, gingivitis, stomatitis and erythrism.[76]

Among the most poisonous species of organic compounds is methylmercury, which can be produced both as a result of industrial processes and through natural means. It was discovered in Japan in 1956 as a result of prolonged discharges from a chemical plant into Minimata Bay, from which much of the population consumed seafood. The facility used mercury sulfate in the production of acetaldehyde, creating methylmercury as a byproduct which was then released in wastewater into the bay. Because seafood was an important part of the local diet, many people became permanently and severely disabled or were subsequently born deformed.[77] The disaster spurred research which ultimately identified methylmercury as the toxin, and also demonstrated that it could also be formed by natural processes. In such cases, methylmercury is produced when mercury deposits in slow moving or still bodies of water and is consumed by anaerobic bacteria. Their metabolic process converts it to methylmercury, which is then excreted and consumed by fish and shellfish where it bio accumulates, or is not excreted as fast as it accretes. Bioaccumulation also occurs in the food chain, and as a result large, long-living fish such as tuna and swordfish commonly have higher levels of methylmercury than their prey. The popularity of such fish in many diets has prompted warnings from public health officials regarding their excessive consumption, especially among pregnant women.[78]

Whatever the species, mercury intoxication has physical and psychological effects on both humans and animals. Some effects are overt, while others are

---

76   Waldron, 91, 101–103; USEPA, *Mercury Study*, 3–23; "Health Effects," 107, 124, 161, 169, 221; Robins, *Mercury, Mining and Empire*,104; Bidstrup, 77–78.

77   D'Itri, 2, 23; Hugh L. Evans, et al., "Behavioral effects of mercury and methylmercury," in *Federation Proceedings*, Vol. III4, No. 9 (August, 1975), 1860, 1865; David Marsh, 74–75; Schutte, 551; Waldron, 102; Robins, *Mercury, Mining and Empire*, 105.

78   D'Itri, 2–3, 47–51, 63–64; Evans, "Mercury," 999; Robins, *Mercury, Mining and Empire*, 106.

subclinical. Physical symptoms are most apparent in cases of acute elemental mercury poisoning, in which the subject is exposed to a significant amount in a limited time. This can cause tremors, nausea, exhaustion, rapid heartbeat, hallucinations, digestive and kidney afflictions, and reduced sensation in their arms and legs.[79] Acute poisoning resulting from the inhalation of mercury vapor, which quickly penetrates the brain, may cause hyper salivation and inflammation of the lungs and airways, resulting in difficulty breathing, suffocation and death. In the case of methylmercury poisoning, after a latent period, those poisoned often suffer from constricted vision, a tingling sensation in the limbs, ataxia, or the severe loss of control over ones movements, difficulty speaking, hearing problems, excessive perspiration, and a staggering walk.[80]

Chronic mercury poisoning results when an organism is exposed to a limited amount of a form of mercury for over ten percent of their life. Symptoms of chronic exposure to metallic and organic mercury often manifest orally, with hyper salivation, swollen gums and salivary glands, halitosis, loose teeth, oral ulcers, a cough and a black gum line as common symptoms. These are often accompanied by a metallic taste, anemia, weight loss, skin disorders, vision problems, photophobia, cramping and a decline in immunity and motor function. In more severe cases, individuals may suffer quivering in their fingers and eyelids, which may progress to severe shaking in their extremities. This may also result in intention tremors in the hands, in which the more a person tries to perform a task, the more their hands shake. Tremors may also progress to ataxia which can prevent them from dressing, walking and eating.[81] In some

---

79   D'ítri, 15, 123, 142, 205; Schutte, 551; "Health Effects," 219; Hugh Evans, "Behavioral effects of mercury and methylmercury," 1865; Iden, "Mercury," 1001; Bidstrup, 8, 42–43, 60; Schutte, 552; Paul Neal, et al. *Mercurialism and Its Control in the Felt-Hat Industry,* (Washington, DC: Government Printing Office, 1941), 122; Waldron, 103; Cobb, "Supply and Transportation," 40; "Health Effects," 31, 33, 47–49, 51, 54–55, 58, 162, 163,185–186, 264–265, 268, 272; USEPA, *Mercury Study,* 0–3; Helena Hanninen, "Behavioral Effects of Occupational Exposure to Mercury and Lead," in *Acta Neurologica Scandinavica,* Vol. VI6, Supplement 92 (1982), 170; Cherry, 79; Piikivi, 35; Robins, *Mercury, Mining and Empire,* 106–107, 110.

80   "Health Effects," 34, 43; Robins, *Mercury, Mining and Empire,* 107; D'Itri, 24–25; Hugh Evans, "Behavioral effects," 1860; Marsh, 74–75.

81   United States Environmental Protection Agency, *A Review of the ReferenceDose and Reference Concentration Processes.* EPA/630/P-02/002F. Risk Assessment Forum (Washington, DC.: EPA, 2002), 4–3; Parés y Franqués, 120, 163, 261, 291; Waldron, 104–105; Menéndez Navarro, 166–167; Bidstrup, 42; Neal, 122; "Health Effects," 44, 56; Leena Piikivi, et al., "Psychological performance and long-term exposure to mercury vapors," in *Scandanavian Journal of Work and Environmental Health,* Vol. Io (1984), 35; Schutte, 553; Goldwater, 268–269; Soleo, L, et al., "Effects of low exposure to inorganic mercury on psychological performance," in *British Journal of Industrial Medicine,* Vol. IV7, No. 2 (1990), 105; Reiko Kishi,

cases, the end of tremors is the beginning of mental deterioration, which can range from dementia to insanity. Generally, oral problems and damage to the kidneys, gastrointestinal tract and respiratory symptoms reflect a greater level of exposure, while chronic, low-level exposures will more likely result in neurological problems.[82]

Exposure to mercury vapor and contaminated dust may also result in subclinical and otherwise diffuse symptoms. For example, a study in which two teens were exposed to mercury vapors in their home over a three month period found that they had problems with memory loss and reasoning a year after exposure stopped.[83] In another study, seventy-six retired Japanese mercury miners displayed memory, motor and reaction-time deficits, along with ill tempers, chronic fatigue and insomnia, eighteen years after they stopped working in the mine. Other symptoms included lower back pain, elevated blood pressure, slurred speech, a cough, inflammation of the gums and lung disease. Many reported suffering many of the same symptoms, such as tremors, irritability, depression and gingivitis, when they were working as miners.[84] Mercury mine

---

Rikuo Doi, et al., "Residual neurobehavioural effects associated with chronic exposure to mercury vapour," in *Occupational and Environmental Medicine* Vol. V1, No. 1 (January, 1994), 35–38, 40; Toyoto Iwata, Mineshi Sakamoto, et al., "Effects of mercury vapor exposure on neuromotor function in Chinese miners and smelters," in *International Archive of Occupational Environmental Health*, Vol. VIIIo, (2007), 381, 386–387, Li Ping, Xinbin Feng, et al., "Mercury exposures and symptoms in smelting workers of artisanal mercury mines in Wuchuan, Guizhou, China," in *Environmental Research*. Vol. Io7 (2008), 108–109; Li Ping, Xinbin Feng, et al. "Mercury exposure in the population from Wuchan mercury mining area, Guizhou, China," in *Science of the Total Environment*. Vol. III95 (2008), 72, 76; Irma West and James Lim, "Mercury Poisoning Among Workers in California's Mercury Mills," in *Journal of Occupational Medicine*. Vol. Io, No. 12 (December, 1968), 698; Barboni Salgueiro, Mirella Telles, Marcelo Fernandes da Costa, et al., "Visual field losses in workers exposed to mercury vapor," in *Environmental Research*, Vol. Io7 (2008): 124, 129; D'Itri, 134–137; Williamson et al., 273–274; Kishi, Reiko, Rikuo Doi, et al. "Residual neurobehavioural effects", 35; Robins, *Mercury, Mining and Empire*, 107, 109.

82   Parés y Franqués, 99, 121, 123, 228, 232–233; Matilla Tascón, Vol. II, 319; Schutte, 552.

83   Yeates and Mortensen, 209–211, 214, 218; Robins, *Mercury, Mining and Empire*, 108. The concentration in the home was an eight-hour time-weighted average of 50 micrograms per cubic meter.

84   Reiko Kishi, Rikuo Doi, et al., 35, 40; Reiko Kishi, et al., "Subjective Symptoms and Neurobehavioral Performances of Ex-Mercury Miners at an Average of 18 years After the Cessation of Chronic Exposure to Mercury Vapor," in *Environmental Research*. Vol. VI2, No. 2, (1993), 293–295; Robins, *Mercury, Mining and Empire*, 108. See also A.M. Williamson, et al., "Occupational Mercury Exposure and its Consequences for Behaviour," in *International Archives of Occupational and Environmental Health*. Vol. Vo (1982): 273–286 and A. Andersen,

workers in California also reported almost identical symptoms.[85] In a study of felt hat industry workers whom had been exposed to a mercuric nitrate solution, researchers noted an unwillingness to obey commands and severe personality changes that alienated the workers from each other and their families. Similarly, compared to a control group, research on female dental workers found that they had a higher subclinical incidence of short term memory loss, obsessive compulsive activity, nervousness, and depression and personality changes, all in the absence of tremors or other effects on motor coordination or mental capacity.[86]

Such studies highlight the fact that mercury's subclinical effects are often psychological and can endure longer than post-exposure physical effects. Neuropsychological effects can range from memory loss, lethargy, indecisiveness and decreased confidence and ability to focus, to more severe clinical problems such as depression, nervousness, obsessive-compulsive disorders, verbal and physical aggression and psychoses.[87] Even in cases of clinical symptoms, diagnosing mercury intoxication can be complicated because many afflictions share the same symptoms. As a result, diagnosis is often delayed, or erroneous,

---

et al. "A neurological and neurophysiological study of chloralkali workers previousy exposed to mercury vapor," in *Acta Neurologica Scandinavica*, Vol. VIII8, No. 6, (December 1993), 427, 431; Toyoto Iwata, Mineshi Sakamoto, et al., "Effects of mercury vapor exposure on neuromotor function in Chinese miners and smelters," in *International Archive of Occupational Environmental Health*, Vol. VIIIo, (2007), 381, 386–387; Li Ping, Xinbin Feng, et al. "Mercury exposures and symptoms in smelting workers of artisanal mercury mines in Wuchuan, Guizhou, China," in *Environmental Research*. Vol. Io7 (2008), 108–109; Li Ping, Xinbin Feng, et al. "Mercury exposure in the population from Wuchan mercury mining area, Guizhou, China," in *Science of the Total Environment*. Vol. III95 (2008), 72, 76; Salgueiro Barboni, Mirella Telles, Marcelo Fernandes da Costa, et al., "Visual field losses in workers exposed to mercury vapor," in *Environmental Research*, Vol. Io7 (2008), 124, 129.

85 West and Lim, 698; Robins, *Mercury, Mining and Empire*, 109.

86 Neal, 84; USEPA, *Mercury Study*, USEPA, *Mercury Study*, 2–6.

87 Parés y Franqués, 294; Waldron, 104–105; Piikivi, "Psychological performance," 35; Schutte, 553; Evans, "Mercury," 1001; Neal, 122; Hanninen, 170; Bidstrup, 42; "Health Effects," 58, 220; L. Soleo, et al. "Effects of low exposure to inorganic mercury on psychological performance," in *British Journal of Industrial Medicine*, Vol. IV7, No. 2 (1990), 105; Richard Ehrenberg, et al., "Effects of Elemental Mercury Exposure at a Thermometer Plant." in *American Journal of Industrial Medicine* Vol. I9 (1991), 500; Yeates and Mortensen, 209–211, 214, 218; Reiko Kishi, et al., "Subjective Symptoms and Neurobehavioral Performances of Ex-Mercury Miners at an Average of 18 years After the Cessation of Chronic Exposure to Mercury Vapor," in *Environmental Research*. Vol. VI2, No. 2, (1993), 293–295; West and Lim, 698; P.J. Smith, 413, 417, 419; Williamson, et al, 273, 284–285; Robins, *Mercury, Mining and Empire*, 107–108.

although examining occupational or residential histories is often helpful in determining mercurialism.[88]

As we have seen, the debilitating, and often lethal, effects of mercury intoxication were recognized by contemporaries; from mitayos all the way to viceroys. Many of those who were poisoned by mercury or otherwise became ill or injured from mining either died from acute exposure or tried to escape the town. Beyond mercury, silicosis, umpé and cave-ins, Indians also suffered from the sharp temperature difference between the hot mine and cold and often blustery exterior which caused respiratory diseases. There was, however, a hospital in Huancavelica to care for them. In 1588 it offered 120 beds although between 1586 and 1608 it did not have a physician or surgeon.[89] Conditions did improve somewhat as a result of Viceroy Montesclaros' 1608 visit to the town, during which time he more than tripled the budget from 2,000 to 6,250 pesos. In addition, he placed the institution under the administration of the San Juan de Dios religious order, whose vocation is to care for the ill and for whom the hospital was then named. In the late seventeenth century, following the death of the developer of the busconil smelter, Lope de Saavedra, in a shipwreck while bound for Spain, his two percent royalty on mercury production was redirected to subsidize the operation of the hospital.[90]

---

88   D'Itri, 107; Piikivi, "Psychological performance," 25; Bidstrup, 43; N.M. Cherry, "Neurotoxic effects of workplace exposures," in *Hunter's Diseases of Occupations*, P.A.B. Raffle et al., Eds. 8th ed. (London: E. Arnold, 1994), 79; P.J. Smith, "Effects of occupational exposure to elemental mercury on short term memory," in British Journal of Industrial Medicine, Vol. IVo (1983), 413, 417; Barbara Uzzell and Jacqueline Oler, "Chronic Low-Level Mercury Exposure and Neuropsychological Functioning," in *Journal of Clinical and Experimental Neuropsychology*, Vol. VIII, No. 5 (1986), 581–582; Robins, *Mercury, Mining and Empire*, 105, 108.

89   Emetherio Ramírez de Arellano, 1; Montesinos, Vol. II, 187; Whitaker, 19; Sala Catala, 196; Brown, "Worker's Health," 478; Lohmann Villena, *Las minas de Huancavelica*, 224; Robins, *Mercury, Mining and Empire*, 137; Juan de Mendoza y Luna, "Carta del Virrey Marqués de Montes Claros a S.M. en materia de Real Hacienda," 90; Cantos de Andrade, 308; Patiño Paúl Ortíz, 63–64; Lohmann Villena, *Las minas de Huancavelica*, 224; Cobb, *Potosí y Huancavelica*, 78; Brown, "Worker's Health," 478; Robins, *Mercury, Mining and Empire*,137. See also Francisco de Borja, 86.

90   Marqués de Casa Concha, Relación del estado que ha tenido y tiene la real mina de Guancavelica, 52–53; Provisión del virrey Diego Fernández de Córdoba, sobre el hospital real de la Villa de Huancavelica. Cuzco, March 21, 1623. ADC, Fondo Corregimiento, Legajo No. 7, No. 114, Cuaderno 3, 1–3; Toledo y Leiva, 183; Navarra y Rocaful, 175; Caravantes, Vol. IV, 176; Patiño Paúl Ortíz, 64–65; Carrasco, 154; Contreras, 49; Lohmann Villena, *Las minas de Huancavelica*, 224, 226–227, 434; Sala Catala, 197; Fernández Alonso, 367; Povea Moreno, *Retrato de una decadencia*, 265–267; Robins, *Mercury, Mining and Empire*, 138.

Further improvements to the hospital came during the administration of Governor José de Santiago-Concha y Salvatierra, Marqués de Casa Concha, from 1723–1726. Not only did he see to it that the hospital received funds in a timely manner, but he also expanded the facility, imposed a degree of long overdue civilian financial oversight, and instituted the raising of sheep on hospital lands to serve as a source of food for the patients. Subsequently, between 1771 and 1776, Governor Domingo Antonio Jáuregui oversaw extensive renovations to the structure. Despite these efforts, the hospital remained a very unhygienic and insalubrious place, although patients did receive a twice daily ration of lamb and potato stew and bread.[91] In the late 1750s, Antonio de Ulloa described the hospital as being filled with "filth... there were not mattresses... there was no pharmacy, nor medicines."[92] Understandably, most of the ill were unwilling to seek care in the hospital, seeing it as nothing more than a place where they would be bled, made to sweat, ministered to, and die. Damián de Jeria, the Protector of the Indians, remarked how the

> doctors... have served very little... it is true what the sick Indians say that they find themselves worse and as a result they flee the hospitals... and sooner or later they all are finished off, with some spitting mouthfuls of blood covered in mercury, which is also sometimes found when they bleed them and also in the graves where they bury them.[93]

Instead, many Indians sought the care of sorcerers, and made offerings to indigenous gods in the hope of recovery.[94] Others sought a warmer climate which not only removed them from the source of the exposure, but also enabled them to literally sweat out their toxins. While this would help, depending on the level of exposure, tremors and psychological effects may be long-term or permanent.[95]

---

91   Marqués de Casa Concha, Relación del estado que ha tenido y tiene la real mina de Guancavelica, 52–54; Fernández Alonso, 366–367; Patiño Paúl Ortíz, 65; Povea Moreno, *Retrato de una decadencia*, 268–269, 273–274.

92   Molina Martínez, *Antonio de Ulloa*, 158.

93   Jeria, 3, see also 5.

94   Basto Girón, 56–57, 73.

95   Basto Girón, 56–57, 73; Montesinos, Vol. II, 187; Ulloa, *Noticias* Americanas, 281; Brown, "Worker's Health," 478, 485–486; Lohmann Villena, *Las Minas de Huancavelica*, 226; Basadre, "El Régimen de la Mita," 348; Molina Martínez, *Antonio de Ulloa*, 89; Sala Catala, "Vida y muerte," 195, 197, 200; Goldwater, 40; Purser, 44; Schutte, 554; Bidstrup, 60; Cherry, 79; Kishi, "Residual neurobehavioural effects," 40; Waldron, 106; Robins, *Mercury, Mining and Empire*, 137.

Excessive alcohol consumption formed a part of the denigratory Hispanic narrative against Indians, with some, such as Friar Reginaldo Lizárraga, arguing that it was at the root of the demographic decline.[96] Ironically, among those who worked with mercury, either in Huancavelica or in silver mills, alcoholic beverages were one of the few things that would limit their body's absorption of elemental mercury vapor. Alcohol and elemental mercury both bind to red blood cells, however ethanol restricts the release of catalase, an enzyme which enables the oxidation of mercury in the blood.[97] As a result, alcohol inhibits both the uptake, and oxidization, of mercury vapor in the lungs by up to fifty percent. There is a downside, however, as mercury which enters the body in these conditions remains there for a longer time and is more likely to end up in the liver.[98]

Although there is a paucity of descriptions of mercury intoxication in Huancavelica, they abound for Spain's mercury mine in Almadén, thanks to the work of José Parés y Franqués who served as physician there from 1761 to 1798. He described workers, often children, suffering from pallidity, weight loss, the shakes, gingivitis, hyper salivation, persistent cough, edema and respiratory problems.[99] Parés y Franqués related how severely ill patients sweated profusely, had to be tied down to limit damage from tremors and experienced severe chest pains and shortness of breath.[100] He also described incapacitating motor effects, noting that "some could not stand, others could not even sit alone, infinite numbers cannot eat or drink without assistance, most walk by skipping or trotting, as they cannot walk in a normal manner."[101]

---

96   Lizárraga, 209; Robins, *Mercury, Mining and* Empire, 15.

97   Ping, et al., "Mercury exposures and symptoms," 112.

98   Ping, et al, "Mercury exposures and symptoms," 112; Kishi, "Residual neurobehavioural effects," 39, 41; Nielson Kudsk, "The Influence of Ethyl Alcohol on the Absorption of Mercury Vapour from the Lungs in Man," in *Acta Pharmacologia et Toxicologia,* Vol. II3 (1965), 264, 268–269; John Hursh, et al., "The Effect of Ethanol on the Fate of Mercury Vapor Inhaled by Man," in *The Journal of Pharmacology and Experimental Therapeutics,* Vol. II14, No.3 (1980), 520, 522, 524–526; John Hursh, et al. "Clearance of Mercury (Hg-197, Hg-203) Vapor Inhaled by Human Subjects," in *Archives of Environmental Health* Vol. II11 (Nov/Dec 1976), 308; Iden, "The Effect of Ethanol," 520; R.A. Kark, "Clinical and neurochemical aspects of inorganic mercury intoxication," in *Handbook of Clinical Neurology.* P. Vinken and G. Bruyn, Eds. Vol. III6 (New York: Elsevier, 1979), 165–166; Robins, *Mercury, Mining and Empire,* 143–144. Concerning rats, see Laszlo Magos, et al. "The Depression of Pulmonary Retention of Mercury Vapor by Ethanol: Identification of the Site of Action," in *Toxicology and Applied Pharmacology,* Vol. II6 (1973): 180–183.

99   Parés y Franqués, 24, 80, 82, 84, 122–123, 261; Robins, *Mercury, Mining and Empire,* 138.

100   Parés y Franqués, 239; Robins, *Mercury, Mining and Empire,* 138.

101   Menéndez Navarro, 155.

In 1774, the superintendent of the Almadén mine, Gaspar Soler, described conditions in the mineshafts there. Despite improvements in ventilation and rotating workers, he described a situation where "at every step I encounter many poor people unable to continue working in the mines, in the flower of their youth, some such shakers that sometimes they cannot stand; others extremely weak, others...needing the help of crutches to stand and not a few who were left insensate and stupid."[102] Similarly, Joaquín Ezquerra del Bayo described the health conditions of workers there in 1839. In so doing, he described how mercury vapor produced "the saddest and most pernicious effects... [a] horrible and shocking illness."[103]

Workers in, and residents of, Huancavelica were not only poisoned by mercury, but were also exposed to arsenic and lead. Ingestion or inhalation of low levels of arsenic may result in reduced production of white and red blood cells, heart arrhythmia, vascular damage, nausea and tingling in the extremities. Chronic exposure to inorganic arsenic can result in skin discoloration and produce eruptions. Unlike mercury, inorganic arsenic compounds appear to be more toxic than organic compounds, and may cause kidney damage and increase cancer risks, especially those of the skin, liver, bladder and lung. Like mercury, arsenic may result in stillbirths and birth deformities, and is carried in mother's milk.[104] The presence of arsenic in the soils and travertine rocks of Huancavelica is hardly surprising, given that the two heavy metals are often found together. What is surprising are the levels of lead found in the soils of Huancavelica. This metal primarily affects the nervous system, brain and kidneys, although it can also be a source of joint pain, increased blood pressure, anemia and decreased sperm production. It can also cause miscarriages, premature births, and lower birth weights, and is believed to be a carcinogen. Infants and children are especially at risk as even low exposure levels may impair cognitive development and result in smaller body size.[105]

---

102   Dobado Gonzalez, Rafael, "Salarios y condiciones," 410–411.

103   Dobado Gonzalez, Rafael, "Salarios y condiciones," 411.

104   United States Environmental Protection Agency, "Arsenic, inorganic," Integrated Risk Assessment System (IRIS), Chemical Assessment Summary (Washington, D.C.: National Center for Environmental Assessment, 1988o, 3–17,) see http://cfpub.epa.gov/ncea/iris/iris_documents/documents/subst/0278_summary.pdf. Se also http://www.atsdr.cdc.gov/toxfaqs/index.asp.

105   Thoms, Robins, Ecos, Brooks and Espinoza Gonzales, Results of June-July, 2015 Field Study, 12; United States Environmental Protection Agency, "Lead and compounds," Integrated Risk Assessment System (IRIS), Chemical Assessment Summary (Washington, D.C.: National Center for Environmental Assessment, 2004), 2, 7–9. See also Tables 3 and 8 in

It was not just humans who were poisoned by this toxic cocktail of heavy metals in Huancavelica, all living things were. Foraging animals, such as llamas, mules, cattle, goats, sheep and pigs were exposed to mercury as a result of breathing, grazing or carrying ore. Dogs, cats and other animals were also exposed, and, like humans, suffer from trembling, oral problems, excessive salivation, spontaneous abortions and birth deformities. Some poisoned animals were consumed by humans, adding to their seemingly ever increasing body burden. Plants are also harmed by mercury, and if they are not killed by it may experience root decay and stunted growth.[106]

Ecological Echoes

Mercury mining in Huancavelica had local, regional and ultimately global ecological effects. Locally, deforestation was a consequence of the need for fuel for the smelters, as well as in mine supports and construction. The prime victim was the queñua tree, which is the only evergreen which thrives above 3,300 meters above sea level. Its generally small size facilitates its harvesting, its highly resinous wood and needles serve as excellent fuels and its strength and durability make it an attractive building material.[107] As smelting got underway in the region of Huancavelica in 1564, it was queñua which fired the furnaces. Over the next decade, Huancavelica was surrounded by an expanding radius of deforestation as most queñua had been harvested in an eighty kilometer radius of the town. Ichu stocks would follow a similar path after Rodrigo de Torres de Navarra demonstrated its efficacy as a smelter fuel in 1572. Fifteen years later, it had disappeared within a six kilometer radius of the town.[108]

Less visible, yet infinitely more toxic and widespread, was the release of elemental mercury vapor in the region. Its fate was initially influenced by

---

      appendix and http://cfpub.epa.gov/ncea/iris/iris_documents/documents/subst/0277_summary.pdf; http://www.who.int/mediacentre/factsheets/fs379/en/.

106   USEPA, *Mercury Study*, 2–6; D'ítri, 15, 123, 142, 205; Cobb, "Supply and Transportation," 40; Schutte, 551; Bidstrup, 34; Lohmann Villena, *Las minas de* Huancavelica, 224; Robins, *Mercury, Mining and Empire*, 109–110.

107   Lohmann Villena, *Las minas de Huancavelica*, 52; Patiño Paúl Ortiz, 190; Robins, *Mercury, Mining and Empire*, 19.

108   Cantos de Andrade, 304; "Confirmación de Su Magestad," 191; NA, "Memorial y relación de las minas de azogue del Pirú," 437–438; Patiño Paúl Ortíz, 190–191; Lohmann Villena, *Las minas de Huancavelica*, 54; Contreras, *La ciudad de mercurio*, 20–21; Llano Zapata, 219; Robins, *Mercury, Mining and Empire*, 33, 178.

the steep valley in which Huancavelica sits, as well as its climate. In some conditions, a thermal inversion would cloak the town in a noxious veil of mercury and sulphur-infused smoke. On other days, prevailing winds propelled it through the valley to deposit in the surrounding countryside. Given that mercury can remain airborne for up to a year, Huancavelica's emissions permeated the atmosphere, contributing to background levels of atmospheric mercury.[109] Seeking to avoid exposure to what they knew were noxious vapors, many of Huancavelica's elite constructed their homes in the Santa Ana neighborhood, where it appears that prevailing winds offered some protection from smelting operations. Today, Santa Ana has the lowest level of contamination among earthen homes in the city.[110] Nevertheless, to varying degrees all of the town's residents were exposed to mercury vapor in the air, and also through ingestion by breathing or consuming contaminated dust and foodstuffs.

Of the mercury which deposited in the town, some of it bound with the soil, much of which remains there to this day. Another portion made its way into the Ichu River as runoff during rainstorms. Sediment analysis has indicated extraordinarily high amounts of mercury there, ranging from .88 mg/kg upstream, to 1,370 mg/kg where the river passes through the city, and 18.93 mg/kg downstream.[111] The Ichu River may also contain hyper-toxic methylmercury, in sections where the current is less and the metal is in an anaerobic environment.[112]

The nature and extent of cinnabar extraction and refining largely determined its human and environmental impact. Although open pit mining, as was done until the early 1600s, was less toxic than shaft mining, it was still a poisonous task. Workers were usually immersed a couple of feet deep in a cinnabar muck, allowing dermal absorption as well as ingestion of dust and mercury vapor. Pneumonia and other respiratory diseases were common as the work took place at over 3,650 meters above sea level in what is often a chilly, windy, and wet environment. While fear of Santa Bárbara's pit inspired many to flee the mita, conditions worsened with the advent of shaft mining. Not only were cave-ins, rockslides, and umpé constant menaces, but the lack of ventilation greatly increased the exposure of workers to mercury vapor and silica. It was

109   USEPA, *Mercury Study*, 0–1, 2–1; Robins, *Mercury, Mining and Empire*, 102.

110   Patiño Paúl Ortíz, 40; Lohmann Villena, *Las minas de Huancavelica*, 224; Hagan, et al., "Residential Mercury Contamination in Adobe Brick Homes," 4, 5, 7, 8.

111   CETOX Laboratory results of samples taken July 3 and July 4, 2015.

112   D'Itri, 49–51; Hugh Evans, "Mercury," 999; Bryn Thoms and Nicholas Robins. "Remedial Investigation. Huancavelica Mercury Remediation Project. Huancavelica, Peru," unpublished Manuscript. July, 2015, 7, 7, 25, 26, 38; Robins, *Mercury, Mining and* Empire, 106.

during this period that conditions were worst in the mines. With the opening of the Bethlehem Adit in 1642 a new era began, one with better ventilation and much greater ease of access for inputs and the carrying out of ore. Improving conditions led to increasing numbers of alquilas at the site, and they formed a majority of workers there by the 1650s.[113]

More dangerous than ore extraction was refining cinnabar. The widely used busconil furnace was still quite inefficient, probably losing at least thirty percent of what they produced. Porous ceramic tubes and containers, poorly sealed joints, overheating and the windy climate all contributed to their less than optimal operation. Workers, and their families, were not just exposed to mercury vapor in the proximity of the ovens, but especially when mitayos were forced to enter the smelters before they cooled, or when their families cleaned the various tubes and made bolas.

There was a wide range of afflictions which mercury poisoned workers, and residents, suffered. These included physical conditions such as the shakes, oral problems, nausea, indigestion, peripheral polyneuropathy and weight loss. The range of psychological problems was no shorter, and included diffidence, psychosis, agitation, hostility, difficulty concentrating, depression and obsessive-compulsive disorders.

Although many of these workers became sick, few wanted to go to Huancavelica's hospital. Most viewed it as a place where they would suffer and die, and preferred to seek help from a native healer or make their way home or to a warmer climate to sweat out the mercury in their bodies. Their misgivings were not not unreasonable; bloodletting was the primary treatment, the hospital was maladministered and undersupplied, and the air was contaminated there as it was throughout the town.

At the close of the colonial period, Huancavelica was already in decline. Humanchis dominated production, many of the mines were closed due to the 1786 cave-in, and commerce was dwindling. With the independence wars, many miners fled to the coast, or Spain, further reducing production and economic activity. While this was bad for business, it also signaled the beginning of an era of much less contamination, although residents would, and do, continue to contend with the legacy effects of about 72,125 metric tons of mercury production.

---

113   Fernández de Castro, "Advertencias que hace el Conde de Lemos," 254. See also Haenke, 136; Fernández de Castro, "El Conde de Lemos da cuenta," 271. Haenke, 136; Lohmann Villena, *Las minas de Huancavelica*, 242, 253, 330–333, 353, 382; Brown, "Worker's Health," 485; Robins, *Mercury, Mining and Empire*, 62.

CHAPTER 6

Desuetude, Decay and Neglect: The Nineteenth and Twentieth Centuries

The wars for independence heralded important changes in Huancavelica, which included the Spanish abolition of the mercury monopoly in 1811, and that of the mita, briefly, in 1812. In 1824, Simón Bolívar passed through the town, although fanfare ultimately led to disappointment. By 1826, the Liberator had downgraded the region from the status of a department, or state, with its capital in Huancavelica, to merely a province of the department of Ayacucho. The cinnabar deposits which had been property of the king became that of the newly formed Peruvian government. The mineshafts lay largely abandoned, however, and Peru became an importer of mercury for much of the nineteenth century. With only humanchis producing quicksilver, and little other economic activity, the town slipped into a state of suspended animation. Such was the decay that a visitor in 1834 described it as having less than 4,000 inhabitants and having been "reduced to misery, due to the lack of working the mines."[1]

The region's pride, if not economy, received a boost in late 1839 as President Augustín Gamarra reestablished the department of Huancavelica. With its namesake city as its capital, the department encompassed the provinces of Huancavelica, Castrovirreyna and Tayacaja. Almost a decade later, the city's population had grown to around 5,000, although following the discovery of mercury in California in 1848, the economy, and population of Huancavelica, entered a renewed period of decline. By 1900, only about 3,000 people lived there, a contraction of around forty percent from 1850.[2]

The role of humanchis in producing mercury in the nineteenth century has, until recently, largely been ignored. This partially results from the fact that, following the abolition of the mercury monopoly, there was no systematic and

---

1 Government of Peru, *Almanaque de Huancavelica*, 19–20; Iden, *Censos Nacionales 1993, IX De Población, IV De Vivienda*, Vol. I, 24; Contreras and Díaz, 3, 5, 8; Crosnier, 43; Arana, 15–16; Berry and Singewald, 24; Hawley, 6; Jiménez, 203; Contreras and Díaz, 4–5, 8–9, 11, 23; Strauss, 561; Carrasco, 291; Blanco, 38.

2 Raimondi, "Huancavelica y mina de azogue, año 1812," in *Notas de viajes para su obra "El Perú."* Vol. III (Lima: Imprenta Torres Aguirre, 1945), 285; Government of Peru, *Almanaque de Huancavelica*, 20; Iden, ru, *Censos Nacionales 1993, IX De Población, IV De Vivienda*, Vol. I, 24; Salas Guevara, *Villa Rica de Oropesa*, 69; Contreras and Díaz, 13.

centralized recording of mercury production in the region until 1903. Like their forebears, humanchis forsook safety for production, often mining pillars and similar structures in the mine. Much of the mercury they produced was dispatched to regional silver producing towns such as Castrovirreyna, Cerro de Paso, Jauja, Lircay and Lucanas. It was not always an easy sell, however. Even with their minimal production and transportation costs, humanchis could do little to compete against the cheap California mercury, which sold in the region for less than that produced in nearby Huancavelica.[3]

The rise of Californian quicksilver put an end to efforts in the 1830s and 1840s to revitalize Santa Bárbara's production. One such effort had been led by Demetrio Olavegoya, who in 1836 established a company to reactivate the mine in association with the administration of Andrés Santa Cruz (1836–1839). This effort to satisfy Peru's domestic quicksilver needs was, however, short lived and lasted only three years. Although the initiative did result in a total production of around 184 metric tons of mercury, this only covered about a third of Peru's needs and had largely sidelined independent producers. With the downfall of the Santa Cruz government came the downfall of the company, and humanchi production again came to the fore.[4] By around 1840, there were about 500 humanchis producing about 138 metric tons of mercury annually, more than double that of Olavegoya's operation. To do so, they not only worked the Santa Bárbara and Chaclacatana hills, but also the deposits of La Trinidad, Quirasquichiqui, Corazon-pata and Botija-Punca, refining the ore in seventy-six ovens dispersed in thirteen different locations.[5]

Following the fall of Santa Cruz, the second administration of Augustín Gamarra (1838–1841) developed a similar scheme to exploit the mines in 1840. Like Olavegoya's enterprise, the Sociedad Mineralúrgica had leading civil and military officials among its shareholders. Seeking quick profits, they engaged in the well-established practice of excavating ore from the structural supports in the mine. A lack of technical expertise and dividends led to frictions among the principals, and ultimately the company dissolved along with the Gamarra government in 1841.[6] In the latter 1840s, during his second term President Ramón Castilla (1845–1851) sponsored the establishment of a new entity, known as the Compañía Huancavelicana, which had among its principals

---

3 Contreras and Díaz, 6, 8, 12–14; Povea Moreno, "Los buscones de metal," 135; Dueñas, 156.

4 Rivero y Ustariz, *Memoria Sobre El Rico Mineral De Azogue*, 5; Gastelumendi, 46; Arana, 17; Strauss, 561; Yates, 19; Umlauff, 17; Contreras and Díaz, 12, 14.

5 Rivero y Ustariz, *Memoria Sobre El Rico Mineral De Azogue*, 23, Contreras and Díaz, 8, 11.

6 Arana, 17; Strauss, 516; Berry and Singewald, 24; Umlauff, 17; Rivero y Ustariz, *Memoria Sobre El Rico Mineral De Azogue*, 5; Contreras and Díaz, 16.

Olavegoya and government officials. Initial hopes were high, although it soon became apparent that the new deposits they encountered could not be worked due to a high arsenic content. This contributed to the failure of the company within five years, aided by a lack of technical expertise, investment, politicized leadership and the rise of Californian quicksilver.[7]

In this new competitive context, the subsequent government of José Rufino Echenique (1851–1855) determined to try a new approach and leased the mine to Luis Flores for a decade for 3,000 pesos a year. Flores' efforts to attract investors were largely fruitless, and the operation remained politicized, with Flores working in reluctant partnership with Juan Salaverry, the prefect of Huancavelica. Flores' operation employed 250 workers, and, like his predecessors, used traditional smelting techniques. By 1854, he had abandoned the effort, having produced a paltry forty-seven metric tons a year for three years. Beyond poor ore, antiquated refining techniques, politicization and competition, the legalization in 1851 of the export of unrefined cinnabar further depressed the demand for mercury. There were also other, continuing, structural problems which inhibited production, such as those associated with Huancavelica's isolated location and a lack of incentives for investment.[8] As in previous occasions, with the shuttering of Flores' operation, humanchi production again returned to prominence and demonstrated that, given the circumstances, they were the more efficient producers. While Flores' 250 employees produced forty-seven metric tons a year, 250 humanchis would produce about sixty-seven.[9]

In 1863, nine years after Flores' abandoned his effort to reactivate the mines, the newly formed Basadre Company proposed a 1,000,000 peso investment in the mine and new smelters in exchange for a fifty year lease, duty free importation of technology, and exemption from military service to attract workers. Because he was not included in the proposal, former governor Salaverry, now a senator for Huancavelica, obstructed it in congress. As a result, the endeavor did not prosper, and humanchi production continued to prevail.[10] Subsequently, the government's call in 1871 for bids to work the mine elicited little response due to the ensuing global economic recession of 1873.

---

7   Arana, 17; Strauss, 516; Berry and Singewald, 25; Carrasco, 276; Rivero y Ustariz, *Memoria Sobre El Rico Mineral De Azogue*, 5; Contreras and Díaz, 16.

8   Berry and Singewald, 25; Umlauff, 46; Rivero y Ustariz, *Memoria Sobre El Rico Mineral De Azogue*, 5; Contreras and Díaz, 17; Contreras and Díaz, 17–18, 23–25; Contreras, "El reemplazo del beneficio de patio," 411.

9   Berry and Singewald, 25; Umlauff, 18; Contreras and Díaz, 18.

10   Umlauff, 18.

By the 1880s, however, economic conditions had improved and León Alarco organized a company which leased the mine from the government under the first Presidency of Andrés Avelino Cáceres (1883–85). No sooner had operations begun, however, than political unrest undermined the project. Following revisions to the mining code in 1877 which allowed private ownership of mines and foreign investment, a project with British capital was floated in 1890. It too was ultimately abandoned amid humanchi resistance and fears that more modern extraction and refining techniques would put them out of business.[11]

The mercury economy in Huancavelica faced an additional problem beyond those of a lack of investment, politicization of production schemes, humanchi opposition to large scale production, and the decreasing price of mercury due to California's production. In the 1850s, quicksilver was increasingly replaced in silver production with the solvent-based lixiviation system. This German invention allowed silver refiners to extract eighty-five percent of the silver contained in the ore, as opposed to sixty percent through mercury amalgamation, with considerably less expense and manpower.

Traveler's accounts provide a glimpse of the town during the mid-nineteenth century. Among them is that of the Italian naturalist Antonio Raimondi who visited Huancavelica in 1862, having been there previously in 1858. In his notes he described, almost minute by minute, his forty-five kilometer trip from Coati along a road which was passable when it was dry, but in the rain quickly turned to a sticky clayish mud, making passage extremely difficult. On the route, he noted the abundance of camelids, and once in Huancavelica, he drafted a map of the town and visited the mines. He described the town as having "not very straight streets... which vary considerably in width... the houses are of good appearance, all with tile roofs and whitewashed walls." The churches of Santo Domingo and Ascención had "their facades, except the towers, painted a bright red, which clashes with the white of the towers and makes it lose much of its attractive appearance." He added that "in the middle of the plaza is found a good stone fountain, constructed in this year [and nearby is the city hall] which has a good arcade on the first floor."[12] The hospital continued to function, alongside a barracks, and a total of seven churches and one school were located in the town, which did not have a printing press.[13]

The prefect Juan Bustamante, who had overseen the installation of the new fountain, had also made improvements to the thermal baths by installing dressing rooms and channeling the water leaving the pool to power

---

11   Contreras, "El reemplazo del beneficio de patio," 401, 407; Contreras and Díaz, 20–21.
12   Raimondi, "Huancavelica y mina de azogue," 286.
13   Raimondi, "Huancavelica y mina de azogue," 286–287; Carrasco, 298.

two mills. Raimondi noted that the town had been "very rich, but since the mercury mines are not being worked, it was getting poorer and poorer."[14] Near the entrance to the Santa Bárbara mine, he described a small population living in stone houses with tile roofs set along a main street. Upon entering the mine via the Belén tunnel, he was able to penetrate only about 450 meters before being forced to crawl, and soon after, retreat, when his candles began to go out.[15]

By the time of Raimondi's third visit to the region in 1866, conditions there had gone from bad to worse. He described the town in "a state of prostration" and "continuous decadence as a result of the almost complete stoppage of work, in the...Santa Bárbara mine." Despite this, like others, he saw the potential of the zone for wool production which could offer "a brilliant future."[16] Another traveler of the time echoed his view, remarking that "since the work in the mines has stopped, [the city] has been decaying more and more every day and soon would have arrived at its complete ruin" without the wool industry. Much of Huancavelica's production of 460 metric tons of wool was exported to Europe, and increased production, the traveler opined, could "bring the department great riches."[17] By the 1890s, little had changed in Huancavelica from one hundred years previously. Visitors described crumbing churches, a town in decay, a derelict mine and humanchis conducting refining operations the close to the pithead using traditional busconil smelters.[18]

Haciendas and Huancavelica

With the decline of Huancavelica's mercury production, the relative economic importance of regional haciendas increased. This is not to suggest, however, that they were the source of much economic dynamism. There are traditionally two types of haciendas in the region of Huancavelica: those above and below 3,500 meters. The higher ones are mostly pastoral, focused on raising camelids and sheep, as well as some cultivation of tubers, quinoa and barley. Below this

---

14   Raimondi, "Huancavelica y mina de azogue," 287.

15   Raimondi, "Huancavelica y mina de azogue," 276–281; Iden, *El Perú*, Vol. I, 162–163.

16   Carranza, "De Huanta á Lima por el camino de Huancavelica – Año de 1866," in *Boletín de la Sociedad Geográfica de Lima*. No. 5 (September 30, 1895), 178; Raimondi, El Peru, Vol. I, 241; Carrasco, 303.

17   L. Carranza, 184.

18   Contreras, "El reemplazo del beneficio de patio," 391, 402, 410–411; Contreras and Díaz, 20–22; Torrico y Mesa, 891–892.

level, haciendas tend to have more cattle and sheep and fewer camelids, and cultivate corn, wheat and barley.[19]

In 1876, within the district of Huancavelica, there were twenty haciendas with a combined population of 2,155 people, the vast majority debt peons. The region's hacendados, and elite generally, had little sympathy for the independence movement, and as it gained momentum, many left for the coast or Spain. Left behind were less prosperous Spaniards and creoles, and a number of economically ascendant mestizos. Many mestizos entered the ranks of hacendados, especially as more land became available in 1827 with the expropriation of the haciendas of religious orders. While these dynamics fostered the emergence of a post-colonial landed oligarchy in much of Peru, in Huancavelica haciendas tended to be fragmented through sale, resulting in a preponderance of medium sized estates.[20]

This trend came to a stop in 1883 with the end of War of the Pacific between Peru, Chile and Bolivia. Like many cities in Peru, Huancavelica was occupied by Chilean forces, and hacendados increasingly resided in the town to escape the rural guerilla war being led by Andrés Avelino Cáceres. Despite the unrest, hacendados ultimately benefitted from these events, prospering as urban merchants and increasing their landholdings, often at the expense of Indian lands. Such was the support of the hacendado class in Huancavelica for the occupying Chileans that they organized a militia to defend them as they withdrew.[21]

Although the hacendado class in Huancavelica became stronger as a result of the War of the Pacific, they would come under new pressures in the early twentieth century. Under the second administration of Augusto Leguía (1919–1930), expanding commercial networks and broader economic growth undermined the trade dominance of Huancavelica's elite, while greater centralization of power in Lima undercut their local political power. This spurred emigration among the elite, and the sale and parcelization of their haciendas. This latter tendency increased beginning in 1959 as fears of imminent land reform spurred many hacendados to sell their holdings. These forces not only led to an increase of Indian community lands, but allowed Huancavelica's mestizos and merchants to purchase smaller estates. Although some managed their holdings directly, more commonly they hired an administrator or simply

---

19   Favre, "Evolución y situación de las haciendas," 247–250.

20   Government of Peru, *Censo general de la República del Perú, Formado en 1876*, 680–681; Henri Favre, "Evolución y situación de las haciendas en la región de Huancavelica, Perú," in *Revista del Museo Nacional*. Vol. III3 (1964), 237; Favre, "Evolución y situación de las haciendas," 240–242.

21   Favre, "Evolución y situación de las haciendas," 242–244.

rented the land to producers. Just as urban Huancavelicanos were purchasing haciendas, many rural peons were becoming urban residents. This demographic shift also resulted from hacendado concerns over control of their lands, and a consequent reorientation of their production towards livestock. This was less labor intensive, and allowed hacendados to remove many peons from their lands, and with them, they hoped, the risk of land seizure. As this process unfolded, it also led to a transition from debt peonage to wage labor among those who remained on haciendas.[22]

Santa Bárbara's Reawakening

Late nineteenth century hacendados were largely alone in their relative prosperity, as the town as a whole continued in a state of decadence. Even humanchi production had declined. Whereas around 1840 they produced about 114 metric tons of mercury a year, by 1909 their annual production had dropped to about eighty-three metric tons. Such conditions prompted one observer to declare in 1905 that "the mercury industry in Huancavelica is dead."[23] The town's 3,000 or so residents lived in homes without piped water or sewerage, although by 1908 communication with the outside world had improved with the installation of a telegraph station. Little had changed by 1922, when visiting American geologists described it as "a backward town of several thousand inhabitants, consisting almost entirely of Indians."[24]

With the issuance of a new mining code in 1901, the government endeavored to stimulate investment and economic growth by opening the mines to new claims. Any reactivation would, however, have to contend with the challenges posed by the rugged isolation of the region. In 1920, there were two ways to get to Huancavelica. One involved taking a day-long steamer trip from Callao to Pisco, followed by a five day mule ride upland on a road used primarily by muleteers supplying the highlands with wine, spirits, sugar and coastal agricultural produce. Somewhat more expedient, especially in terms of importing machinery, was to take the two day train ride from Lima to Huancayo via La Oroya, which left only a two day mule ride to Huancavelica.[25]

---

22 Favre, "Evolución y situación de las haciendas," 244–250, 254–256.
23 Salas Guevara, *Historia de Huancavelica*, 198; Contreras and Díaz, 5; Contreras and Díaz, 12–13; Loveday, 83–84, Hadley, 44.
24 Dueñas, 143–144; Strauss, 561; Berry and Singewald, 15.
25 Berry and Singewald, 16, 25; Contreras and Díaz, 4, 22; Singewald, "The Huancavelica Mercury Deposits, Peru." 518; Arana, 16; Dueñas, 143; Strauss, 561, 566.

Sensing opportunity in the new mining code, and undeterred by the technical and logistical obstacles, the Peruvian Augusto Benavides systematically purchased the Santa Bárbara mount and surrounding lands between 1901 and 1915. Towards the end of 1915, as mercury prices increased with the advent of World War One, Benavides sold his holdings to Eulogio Fernandini, inaugurating a new era in Huancavelica's mining, and environmental, history. The scale of the investment necessary to rehabilitate the mine is shown by the fact that when Fernandini purchased it, the Bethlehem adit was only passable for about 280 meters, and by 1922 the mine was still not producing.[26] Fernandini's plans included the construction of a new tunnel which was to run about 1,250 meters and terminate under the main cinnabar deposit, where it would link with the Belén tunnel. To power the necessary machinery, in 1918 Fernandini oversaw the construction of a hydroelectric plant in the Sacsamarca valley, and by the next year workers were working two shifts, twenty-two hours a day, on the tunnel. Despite this, progress was slow, advancing only about a meter or so a day before he shaft was finally completed in 1926.[27]

Once extracted, ore was ferried to smelters by a cable car system where it would be refined in two Gould rotary furnaces, as were used in California. Standing six meters high and fueled by taquia, each could process a 3,175 kilogram load of ore in about ten hours, depending on its quality. Experiments on site indicated an emissions factor, or loss during refining, of between seventeen to twenty percent, which was expected with such furnaces and an improvement over the traditional busconil smelters. In 1938, Eulogio Fernandini died, and his sons inherited the Santa Bárbara mine. In the coming years, cave-ins, such as those in the Bethlehem and Fernandini adits, limited production, which by 1950 had all but stopped.[28]

Fernandini's initiative was the largest, most complex and capital intensive effort to reinvigorate industrial scale mercury production in Huancavelica since the colonial period. Despite its mixed results, it did pave the way for future exploitation of the mines. On May 7, 1956, Fernandini's descendants reorganized their company, Negociación Minera Fernandini Clotet Hermanos, as the Sociedad Minera Brocal. This enterprise mined cinnabar from 1968 to 1975, mostly excavating ores containing about two to three percent mercury from

---

26   Berry and Singewald, 26.

27   Gastelumendi, 47; Salas Guevara, *Historia de Huancavelica*, 198; Berry and Singewald, 27; Yates, 21; Jiménez, 205. Salas Guevara, *Historia de Huancavelica*, 198.

28   Berry and Singewald, 27; Jiménez, 206; Wise and Féraud, 21–22; Contreras and Díaz, 22; Salas Guevara, *Historia de Huancavelica*, 130; Yates, 1, 21, 26.

the open pits of Chacllatacana and San Roque.[29] Unlike Fernandini's smelters which could process perhaps seven tons of ore in a twenty-four hour period, those utilized by Brocal could refine about 250 metric tons of good ore in the same period, also with rotary smelters. For lesser quality ores, a flotation process concentrated the cinnabar into a higher grade which was then smelted, allowing the processing of about 120 metric tons of such ore a day. In 1971, access to underground deposits was aided by the reopening of the Bethlehem adit, however Brocal would cease production of mercury in 1975.[30]

Mining of other elements such as copper, lead, zinc, gold, and silver also shaped people's relationship with their environment in the department of Huancavelica in the early and mid-twentieth century. In the 1960s, however, small-scale production came under pressure as the ownership of mines became more concentrated and production became increasingly capital intensive and technologically complex, especially in the higher-elevation regions of Castrovirreyna, Lircay and Huancavelica. In the face of these changes, small scale miners and companies were largely displaced.[31]

The new generation of mining enterprises were generally of two types: national and international. Peruvian companies generally had limited technology and concentrated on copper, lead and zinc production, for which the market was volatile. Transnational companies tended to have more capital and technology and focused on gold and silver, for which prices were generally more stable. From the 1960s to the early twenty-first century in the region of Huancavelica, such companies tended to utilize a mobile and often seasonally oriented, rural workforce, as well as semi-skilled people from the city of Huancavelica. In the 1960s, among the rural workers, over ninety percent were drawn from about a dozen, generally pastoral, villages in the region.[32] Most were men who came to work in mining to generate funds for a specific objective, such as building a house, purchasing land or opening a small business. As a result, mine work was often seen as a temporary, even seasonal, endeavor, which also allowed them to maintain their links to, and lands in, their home communities. In addition, their increased income enabled them elevate their social status as sponsors of community rites. As always, mining is a double edged sword.

---

29   Salas Guevara, *Historia de Huancavelica*, 208, 255; Wise and Féraud, 22–23; Carrasco, 534.

30   Gastelumendi, 64; Carrasco, 534; Wise and Féraud, 22; Salas, Guevara, *Historia De Huancavelica*, 255; Cooke et al., "Over three millennia of mercury pollution," 8832.

31   Favre, "La industria minera de Huancavelica," 85.

32   Favre, "La industria minera de Huancavelica," 86. Regarding labor in Cerro de Pasco in the latter nineteenth century, see José Deustua, *The Bewitchment of Silver: The Social Economy of Mining in Nineteenth-Century Peru* (Athens: Ohio University Press, 2000), 83–84.

Grueling work, familial separation and occupational illnesses such as silicosis are the prices these workers pay for a more prosperous life. The benefits of mining are further tempered by the environmental consequences of their activities, especially in terms of contamination of water sources.[33]

High Hopes

Beyond efforts to reactivate cinnabar mining in Huancavelica, economic hopes in the early twentieth century were raised with the construction of the 128 kilometer railroad which links the city with that of Huancayo. With the Huancavelicano Celestino Manchego Muñoz serving as Minister of Development, and advocate for the project, construction began fitfully in 1908. After repeated interruptions, work resumed at a steady pace in 1919. As with mercury production during the colonial era, construction of the railroad utilized draft labor, as a result of a law passed in June, 1920, which subjected all men between eighteen and sixty years old to work on road, rail, bridge, drainage and water supply projects in the country. Unlike the mita, the length of service depended on one's age, but ranged from six to twelve days and as a result was not as onerous, or toxic, as the mining mita. Like the mita, however, people could buy their way out of it if they paid for the equivalent in local wages, or by sending someone in their stead, and could be exempt if they were infirm.[34]

The construction of the railroad became a rallying cry for Huancavelica. Many believed that it would lead to economic growth and better integration with the country through importing machinery and other goods and exporting the region's minerals and agricultural and livestock products. Although the railroad route follows the river valleys of the Mantaro and Ichu rivers, the geography required the construction of seventeen steel viaducts and thirty-nine tunnels. On October 24, 1926, with great fanfare, almost all of Huancavelica's 4,000 residents attended the inauguration of the railroad. Not only were they proud of the engineering accomplishment of creating one of the world's highest rail lines, but they were also proud that it had been constructed with Peruvian capital.[35] Commonly referred to as the *"Tren Macho,"* or Macho Train, it

---

33   Favre, "La industria minera de Huancavelica," 87–89; Bonilla and Salazar, 5, 9, 10, 11, 16, 19,
       21, 25, 26, 31–32; López Cisneros, 46.

34   Delgado de Castro, 25, 28; Carrasco, 468; Mendoza Ruíz, 24–26.

35   Delgado de Castro, 15, 25, 27–28. For a detailed examination of the politicial and finan-
       cial issues, and hopes, regarding the construction of the railroad, see Miguel Pinto Hua-
       racha, and Alejando Sánchez Salinas, *Las Rutas Del Café Y El Trigo: Los Ferrocarriles De*

is often delayed due to mechanical issues or blocked tracks, and, as the saying goes, "it leaves when it wants and arrives when it can."[36]

While the train provides an important freight and passenger service between Huancayo and Huancavelica, it did not bring the golden era that so many had hoped. Perhaps had the rail continued on to Ayacucho as was originally planned, the benefits would have been greater. It did, however, aid in the export of local products, especially before the road to Huancayo was paved. By 1936 the department had become Peru's wheat basket, providing more of the grain for the nation than any other region.[37]

Despite greater national integration and somewhat increased commercial activity, little else changed in the region and it remained mired in poverty. In the 1940s, as in the 1740s, corn, wheat, barley, quinoa and tubers were the primary agricultural products, complemented by livestock products from llamas, alpacas and cattle. Beyond this, the region produced few other non-mineral goods. Humanchis continued to produce a limited amount of mercury, and although the department exported gold, silver, lead, zinc, antinomy and salt, in 1940 the city of Huancavelica did not have a branch of a national bank.[38] At this time, 8,139 people lived there, distributed in 1,735 homes. Although the city's residents included seventy-two foreigners, including Germans, Arabs, Chinese, Spaniards and Slavs, Quechua-speaking Indians accounted for more than three-quarters of the provincial population. The social and health conditions in the region, and city, were abysmal. For example, in 1940 ninety-five percent of the city's residents had no access to potable water, ninety-three percent had no sewerage, and eighty percent had no electricity. Moreover, from 1942 to 1945, the infant mortality level, measured by death before completing their first year, increased from 123.4 per 1,000 births to 153.6.[39] Province-wide, eighty percent of children between the ages of six and fourteen had no formal schooling, eight percent of all inhabitants were blind, almost twelve percent were demented and twenty-one percent were handicapped. The installation of a potable water system in the city, beginning in 1952, did reduce the population's exposure to waterborne illnesses. Initially, the water was piped in from

---

      Chanchamayo Y Huancavelica 1886–1932 (Lima: Seminario de Historia Rural Andina, Universidad Nacional Mayor de San Marcos, 2009).

36   Espinoza Flores, 136; Amador Mendoza Ruíz, *Crónicas Del Tren Macho* (Lima, Perú: Niger Editions, 1998), 23–24; Carrasco, 471.

37   Rivas Berrocal, 21.

38   Government of Peru, Censo nacional de población y ocupación, 1940, Vol. VI, vii–viii, 31.

39   Government of Peru, Censo nacional de población y ocupación, 1940, Vol. VI, vii–viii, 4, 17, 19, 51–52; Government of Peru. *Almanaque De Huancavelica*, 83.

the Calqui Grande River to the treatment facility in Huancavelica, and thence to homes and businesses. In 1979, seeking to keep up with the growth of the city, the system was expanded by drawing directly from the Ichu river. Today, residents also obtain water from the Punco Huaycco spring in the town of Conayca.[40]

The population, if not prosperity, of the city of Huancavelica continued to increase throughout the 1950s. By 1961, the city had a population of 11,039 people, with an additional 439 living at the pithead in Santa Bárbara. Illiteracy continued to plague the region, with seventy percent of those over fifteen years old unable to read or write. The continuing connections of people to the land and waterways of the region was evidenced by the fact that two-thirds of the department's economically active population derived their sustenance directly from their environment, through cultivation, pastoral activities, fishing and hunting. While the increasing global demand for cotton in the 1960s led to its cultivation in temperate zones of the department, the region otherwise continued to produce its traditional products such as corn, wheat, barley and various tubers, legumes and some sugarcane.[41] Other regional developments included the construction of the Tablachaca dam and the Santiago Antúnez de Mayolo hydroelectric plant on the Mantaro River in Tayacaja province in 1962. Designed by the plant's namesake in 1943, the dam is eighty meters high and provides electricity to Ica, Pisco and Lima on the coast as well as the Marconi mining facility. The 1960s also saw an increase of minerals production in the department, as renewed mercury production added to that of silver and gold in the region.[42]

Among the major developments of the late 1960s was the implementation of a land reform program between 1969 and 1971 which sought to break the economic power of the landed elite. In Huancavelica, of 700,000 hectares subject to distribution, only 200,000 hectares from 213 former haciendas were ultimately distributed in the form of cooperative groups. These were largely unsuccessful, in part because, like the mining gremio, the peasantry did not hold their productive assets as individual property, which restricted access to credit and discouraged investment. In addition, two-thirds of the cooperatives

---

40   Government of Peru, Censo nacional de población y ocupación, 1940, Vol. VI, 22, 46–47; Government of Peru, *Inventario Y Evaluación De Los Recursos Naturales De La Zona Altoandina Del Perú*, 323; Máximo Enrique Ecos Lima, personal communication, 3/25/15.

41   Government of Peru, *Centros Poblados: Sexto Censo Nacional De Población, Primer Censo Nacional De Vivienda*, 2 De Julio De 1961, Vol. II, 260, 264.

42   Carrasco, 554–555; Government of Peru, *Centros Poblados: Sexto Censo Nacional De Población, Primer Censo Nacional De Vivienda*, 2 De Julio De 1961, Vol. II, 260.

were under two hectares in size, limiting their productive potential. Despite their low productivity, by 1982 thirty percent of the arable land in the department was held as cooperatives, while fifteen percent was privately owned, and fifty-five percent was held as indigenous community lands.[43]

The 1980s were a difficult time in Huancavelica due to severe economic problems and the rise of the Shining Path guerrilla group, which had an especially severe impact in the regions of Ayacucho and Huancavelica.[44] Beyond fear, violence and death, the results of the war in the region included out-migration and declining investment. Even before the outbreak of the insurgency, economic and demographic conditions in the 1970s were largely stagnant. During that decade, the population had only increased by 15,000 people, making it the slowest growing department in Peru. The rise of the Shining Path precipitated a net out-migration, which in 1981 alone reached 125,000 people. Infant mortality also remained high in the department, at 10.9 percent between 1986 and 1996.[45] Also alarming were rates of maternal death during childbirth. In 1996, at 400 per 100,000 live births, it was more than ten times that of Lima. Not all of the news was bad in the 1990s, however. In 1993, as the population of the city of Huancavelica reached 31,068 people, homes were increasingly integrated into potable water and sewerage systems, and more streets were paved.[46] These are important developments in terms of public and environmental health, especially given the mercury, lead and arsenic which contaminate the soils, and hence dirt streets, in the city.[47]

At the close of the twentieth century, the region continued to reflect its indigenous and Hispanic heritage, with two-thirds of the population speaking Quechua as their first language, and eighty-three percent indicating they were Roman Catholic. Illiteracy has continued to afflict the region, although in 1993

---

43   Government of Peru, *Censo General De La República Del Perú, Formado En 1876*. Vol. IV,
      680–681; Carrasco, 462, 562; Rivas Berrocal, 51–52; López Cisneros, 30.

44   Tesania Velázquez, 68.

45   Government of Peru, *Inventario y evaluación de los recursos naturales*, Vol. I, 1; López
      Cisneros, 11; Government of Peru, *Población, Mujer Y Salud. Resultados de la encuesta de-
      mográfica y de salud familiar, 1996* (Lima: Instituto Nacional de Estadística e Informática,
      1997), 17, 20, 58.

46   Segundo Seclén Santisteban, "Pobreza e inequidad en salud," in *Coyuntura: Análisis
      Económico y Social de* Actualidad. Vol. IV, No. 21 (2013), 10; Government of Peru, *Censos
      Nacionales 1993, IX De Población, IV De Vivienda*, Vol. I, 154; Salas Guevara, *Historia de
      Huancavelica*, Vol. II, 332.

47   Thoms and Robins, 7; Thoms, Robins, Ecos, Brooks and Espinoza Gonzales, *Results of
      June-July, 2015 Field Study*, 6, 11–12; Thoms, Robins, Ecos and Brooks. *Results of June/July
      2016 Assessment of Soil and Fish*, 9–10. See Tables 1, 3, 6 and 7 in appendix.

it was thirty-four percent, just about half of what it was in 1961. Nevertheless, illiteracy in Huancavelica was still almost three times the national average. The gender gap was extreme, with sixty-one percent of females being illiterate versus twenty-five percent among men.[48] Within the department in 1996, as has historically been the case, agriculture and mining of silver, copper, lead and zinc continued to be the dominant means of making a living. Respectively, they accounted for twenty-four and almost twenty percent of occupations. Construction was also an important activity, with almost a third of the economically active population working in that sector. During this time, almost a quarter of urban homes were female headed, and the average household had five people.[49] Overall, the twentieth century witnessed an increasing population in the department, despite out migration in search of employment or to escape the violence of the Shining Path. Whereas in 1940 the departmental population was about 265,000, by 1993, a year after the capture of the guerilla leader Abimael Guzmán, it stood at 400,000. Despite advances in public works and health indices, the department remains the most impoverished in Peru, and the city of Huancavelica is among the most mercury contaminated urban areas on earth.[50]

48 Government of Peru, *Compendio estadístco* departmental, 300; Government of Peru, *Censos Nacionales 1993, IX De Población, IV De Vivienda*, Vol. 1, 38–39.

49 Government of Peru, *Población, Mujer Y Salud. Resultados de la encuesta demográfica y de salud familiar, 1996* (Lima: Instituto Nacional de Estadística e Informática, 1997), 7–8, 13; Iden, *Compendio Estadístico Departamental* (Lima: Instituto Nacional de Estadística e Informática, 1998), 320.

50 Government of Peru, *Censos Nacionales 1993, IX De Población, IV De Vivienda*, Vol. 1, 29; Government of Peru, *Almanaque de Huancavelica*, 94; Government of Peru, *Censos Nacionales 2007, XI De Población Y VI De Vivienda: Resultados Definitivos*. (Lima: Instituto Nacional de Estadística e Informática, 2008), 941; Government of Peru, Instituto Nacional de Estadística, Sistema de Consulta de Principales Indicadores de Pobreza, Mapa de Pobreza, http://censos.inei.gob.pe/Censos2007/Pobreza/; López Cisneros, 11; Robins, et al., "Estimations of Historical Atmospheric Mercury Concentrations from Mercury Refining," 152; Hagan, et al., "Residential Mercury Contamination in Adobe Brick Homes," 1.

An Invisible Legacy: The Twenty-first Century and Beyond

In the early twenty-first century, visitors to Huancavelica encounter a pleasant, compact city of about 49,000 inhabitants situated in the narrow Ichu River valley. Its five historic neighborhoods of Santa Bárbara, Yananaco, Santa Ana, Ascención, and San Cristóbal retain their distinct identities radiating from their individual plazas graced by colonial churches. The central plaza, in the Santa Bárbara neighborhood, is flanked by the city hall and cathedral. Closed to vehicular traffic, the soft murmur of human voices has replaced the sound of automobile exhausts. Although increasingly few in number, robust colonial travertine homes with grand entrances still embellish the architecture of the city. The Ichu River, spanned by six bridges linking the neighborhoods, is bordered on the south side by an esplanade which is the scene of an expansive and colorful Sunday market. Throughout the city, streets are increasingly being paved and sidewalks installed, public works which are especially important in reducing the amount of toxic dust mobilized by vehicular traffic.

Urban growth continues to expand the city's boundaries, with many homes of recent migrants being constructed of contaminated adobe, often precariously perched on the steep inclines which define the valley. The city's residents retain strong connections to their cultural and environmental heritage. Quechua is widely spoken, many families retain agricultural or pastoral lands in the region and continue to practice both ancestral, and Catholic, rites. The climate remains notoriously tempestuous, and a warm, sunny day can quickly be eclipsed by clouds, a flooding rain or battering hail. A twenty minute drive will bring one up to the Santa Bárbara mines, where a colonial church stands, along with the ruins of buildings, the blocked entrance to the Bethlehem adit, and the abandoned workings of the twentieth century mining and refining operations.

Beyond its attractive, hospitable and historical nature, there is another side to Huancavelica; one characterized by extreme poverty, poor health, multiple heavy metal contamination and the tragic memories many retain of the Shining Path. In 2002, an astounding eighty-eight percent of the department's population lived in poverty, and almost seventy-five percent in extreme poverty,

© KONINKLIJKE BRILL NV, LEIDEN, 2017 | DOI 10.1163/9789004343795_009

the highest percent in Peru.[1] Department-wide in 2000, almost half of the urban population had no potable water in their home, and sixty-two percent had dirt floors. As usual, rural conditions were worse, with sixty-four percent of such residents lacking potable water, eighty-eight percent having dirt floors, and almost three-quarters living without electricity.[2] Agriculture continues to be an economic pillar in rural areas, and in 2007 tubers were the dominant crop of the department, accounting for twenty-seven percent of cultivated land and about three and one-half percent of Peru's potato production. Prospecting by various mining companies for copper and gold suggest that increased production of these minerals may be on the horizon.[3]

The depth and breadth of the challenges facing Huancavelica are highlighted when one compares the region with the rest of Peru. For example, while twenty-three percent of Peru's population is engaged in agriculture or pastoral activities, the rate is almost sixty-five percent in Huancavelica. Earthen homes account for one-third of the total in Peru, but in the department of Huancavelica almost eighty-seven percent are earthen. In the city of Huancavelica, where about half of the homes are made of local soils, this presents a major public health issue as many have been shown to contain lead, arsenic and mercury. For example, in a sample of adobe from seventeen homes distributed throughout the city of Huancavelica, arsenic concentrations in all samples were above the background concentration of sixteen parts per million, with concentrations ranging from fifty-one to 1,611 parts per million. Similarly, eleven of the seventeen homes had lead concentrations 300% higher than the U.S. EPA's regional screening level of 400 parts per million for residential exposures. Other research based on interior samples of sixty adobe homes demonstrated that almost a quarter emanated mercury vapor from their walls above screening levels.[4] While almost fifty-five percent of homes in Peru have potable water

---

1   Government of Peru, *Almanaque de Huancavelica*, 94; Government of Peru, *Censos Nacionales 2007, XI De Población Y VI De Vivienda: Resultados* Definitivos, 941; Government of Peru, Instituto Nacional de Estadística, Sistema de Consulta de Principales Indicadores de Pobreza, Mapa de Pobreza, http://censos.inei.gob.pe/Censos2007/Pobreza/.

2   Government of Peru, *Encuesta Demográfica Y De Salud Familiar 2000*: Huancavelica. Vol. VIII, 37.

3   Salas Guevara, *Historia de Huancavelica*, 126, 328.

4   Government of Peru, *Censos Nacionales 2007: XI de Población y VI de Vivienda. Principales Indicadores Demográficos, Sociales Y Económicos a Nivel Provincial Y Distrital Huancavelica*, 13; Thoms and Robins, 7; Thoms, Robins, Ecos, Brooks and Espinoza Gonzales, *Results of June–July, 2015 Field Study*, 6, 11–12; Thoms, Robins, Ecos and Brooks. *Results of June/July 2016 Assessment of Soil and Fish*, 9–10; Government of Peru, *Censos Nacionales 2007: XI de Población y VI de Vivienda. Sistema de consulta de resultados censales. Cuadros estadistcos.* http://censos .inei.gob.pe/cpv2007/tabulados/#. See Tables 2, 5, 6 and 8 in appendix.

and forty-eight percent have sewerage, in the department of Huancavelica the figures are, respectively, twenty-two and eleven percent. Finally, thirty percent of people in Peru cook with wood, although in the department of Huancavelica almost seventy-percent do.[5]

Health and educational indices are no less encouraging. Among children under five years old in 2000, the department had the dubious distinction of having the highest level of severe malnutrition, at twenty-two percent.[6] In 2006, fifty-six percent of the department's population suffered from malnutrition and one-third of the women were illiterate, the highest rate in Peru.[7] Among both men and women at this time, the twenty percent illiteracy rate was almost three times Peru's national average.[8] Further illustrating the dismal social conditions of the department are surveys concerning marital relations. In 2000, a third of married women reported having been physically abused by their partners, with almost seventeen percent reporting frequent abuse.[9] Huancavelica also has the nation's highest rate of anemia among pregnant women, at fifty-three percent in 2012, almost double the national average. Maternal anemia often results in stillbirths, as well as premature and underweight births.[10]

Such macabre statistics reflect the degree to which the city and region of Huancavelica have been excluded from Peru's wider social, and economic, advances. To some degree, this may be related to the so-called "mita effect," which posits that areas subject to the mine labor draft in Peru and Bolivia are characterized by higher levels of poverty and subsistence agriculture, worse health outcomes and lower levels of infrastructural development than areas

5   Government of Peru, *Censos Nacionales 2007: XI de Población y VI de Vivienda. Principales Indicadores Demográficos, Sociales Y Económicos a Nivel Provincial Y Distrital Huancavelica*, 14, 17.

6   Government of Peru, *Encuesta Demográfica Y De Salud Familiar 2000*: Huancavelica. Vol. VIII, 180–181, 195–198, 203.

7   Tesania Velázquez, 68.

8   Government of Peru, *Censos Nacionales 2007: XI de Población y VI de Vivienda. Principales Indicadores Demográficos, Sociales Y Económicos a Nivel Provincial Y Distrital Huancavelica* (Lima: Instituto Nacional de Estadística e Información, 2009), 11.

9   Government of Peru, *Encuesta Demográfica Y De Salud Familiar 2000*: Huancavelica. Vol. VIII, 180–181, 195–198, 203.

10   Oscar Munares-García, Guillermo Gómez-Guizado, and Juan Barboza-Del Carpio, "Niveles de hemoglobina en gestantes atendidas en establecimientos del Ministerio de Salud del Perú, 2011," in *Revista Peruana de Medicina Experimental y Salud Publica*, Vol. II9, No. 3 (2012), 329; Gustavo F. Gonzales, Vilma Tapia, Manuel Gasco and Carlos Carillo, "Hemoglobina maternal en el Perú: Diferencias regionales y su asociación con resultados adversos perinatales," in *Revista Peruana de Medicina Experimental y Salud Pública*. Vol. II8, No. 3 (2011), 485.

not subject to the mita.[11] While research suggests that flight from the mita had a generationally compounding effect in reducing human capital and accumulation in the region, the toxic effects of legacy mercury production may exacerbate the situation.[12]

Not only has Huancavelica's population been neglected, but much of the region's natural resources are underutilized. For example, in 1989 only twenty percent of the arable land in the department was cultivated. Subsistence agriculture characterizes the region, with farmers commonly utilizing traditional tools. Livestock production continues to consist mostly of cattle, sheep, camelids, pigs and goats, producing meat, milk, cheese, butter, wool and hides as byproducts. The region also produces alpaca and vicuña wool, exporting it nationally and internationally, although most camelid meat is consumed locally.[13] Fish, especially trout, complements local diets, either imported from the coast, or raised in fish farms located in high-altitude lagoons or fed by local rivers. With improved access to credit and agricultural extension, the department's microclimates could increase production of many agricultural products, in addition to wool from camelids, which are a valuable, and renewable, economic resource.[14]

Huancavelicanos have, however, received little benefit from what nature offers, and instead the city's residents contend with invisible toxins such as mercury, lead and arsenic which lace the city's soils, and adobe homes.[15] Although the allocation of mining royalties which accrue to the region has fostered public works, this, and national government investment, have done little to alleviate the abysmal social, economic and public health conditions in the region. While the department derives some economic, if not environmental, benefit from mining, this is not the case with the department's hydraulic resources, the scarcity of which will increase with climate change. The water from Huancavelica's watershed irrigates coastal agriculture, and the hydroelectricity

---

11   Dell, 16–17, 33, 44.

12   Dell 32.

13   Rivas Berrocal, 20; Government of Peru, *Inventario Y Evaluación De Los Recursos Naturales De La Zona Altoandina Del Perú*, Vol. II, 375–376, 379; Government of Peru, *Inventario Y Evaluación De Los Recursos Naturales De La Zona Altoandina Del Perú*, Vol. II, 381, 38, 468–469; Zubilete, "En chakus logran más de una tonelada de fibra de vicuña; Dueñas, 148."

14   Government of Peru, *Compendio Estadístico Departamental* (Lima: Instituto Nacional de Estadística e Informática, 1998), 319; Rivas Berrocal, 21–22, 24, 66.

15   Thoms and Robins, 7; Thoms, Robins, Ecos, Brooks and Espinoza Gonzales, *Results of June-July, 2015 Field Study*, 6, 11–12; Thoms, Robins, Ecos and Brooks. *Results of June/July 2016 Assessment of Soil and Fish*, 9–10. See Tables 1, 2, 3, 6 and 8 in appendix.

generated by the Mantaro River illuminates Lima, Ica and Pisco, while many rural Huancavelicanos are literally left in the dark. Indeed, between 1968 and 1980, ninety-three percent of public investment in the region was dedicated to the construction of the hydroelectric plant. Beyond creating limited employment, the project did little to improve the conditions in which people live in the region.

Besides being the poorest capital in Peru, recent studies have demonstrated that Huancavelica is among the world's most mercury-contaminated cities.[16] In 2009, researchers collected soil samples from fifteen sites on three transects in the city. Surface sample total mercury concentrations ranged from 0.1 milligram per kilogram (mg/kg) up to a remarkable 1,201 mg/kg. At the three centimeter depth below the surface, the range was from 2.5 to 688 mg/kg, while the nine centimeter depth ranged from 1.5 to 90 mg/kg. That deeper soils appear to be less contaminated suggests that the surface level contamination derives from tailings or air deposition. Such concentrations pose health hazards through ingestion or dermal absorption.[17]

The situation is, however, exponentially compounded by the fact that about half of the homes in Huancavelica are constructed of local soils, either as adobe or rammed earth structures.[18] Of these, many have unfinished walls and most have dirt floors. A study in 2011 sampled the interior adobe, air, dust and soil of sixty homes distributed in four neighborhoods in the city. Of these, seventy-five percent had total mercury concentrations which exceeded at least one public health screening level. Almost a quarter had mercury vapor concentrations above 1 microgram per cubic meter ($\mu g/m^3$), which is considered by the United States Environmental Protection Agency as an "action level" which calls for the relocation of residents. Beyond this, forty percent had elemental mercury vapor in their homes above the World Health Organization's reference concentration of 0.2 $\mu g/m^3$. Although vapor readings are in a constant state of flux due to such factors as air movement and temperature, the fact that interior walls of residences emit any mercury vapor is cause for concern.[19]

---

16   Robins, et al., "Estimations of Historical Atmospheric Mercury Concentrations from Mercury Refining," 152; Hagan, et al., "Residential Mercury Contamination in Adobe Brick Homes," 1.

17   Thoms and Robins, 20. Samples were drawn from the surface, and from one and three inch depts.

18   Government of Peru, *Censos Nacionales 2007: XI de Población y VI de Vivienda. Sistema de consulta de resultados censales. Cuadros estadístcos.*

19   Thoms and Robins, 6, 23. See Tables 2 and 5 in appendix.

This is all the more the case when such data is extrapolated to the city-wide level. Based on the above studies, it appears that about three-quarters of the city's adobe homes are above at least one screening level for mercury. With just over half of the homes of the city being made of adobe, this suggests that about 19,000 people are potentially exposed to some form of mercury in their home in a concentration which can cause health problems.[20] Most of the families who live in adobe homes are the poorest of the poor, often lacking electricity, sewerage and gas for cooking. Many can be found on Sundays, standing knee deep in the Ichu River doing their laundry. Unfortunately, the city lacks a functioning sewerage treatment plant, and as a result raw waste from the city, and hospitals, is constantly discharged into the river.[21] Such deposits only add to the burden of contamination in the Ichu River, the sediment of which has been shown to contain total mercury concentrations up to 1,370 mg/kg, raising the possibility that some of this has been converted to ultra-toxic methylmercury.[22]

Other, less well-researched, risks to residents in the city are arsenic and lead poisoning. Like mercury, arsenic and lead have been detected in earthen and travertine homes and at levels which present a risk to residents, especially women and children.[23] The presence of arsenic is unsurprising, as this

20   Government of Peru, *Censos Nacionales 2007, XI De Población Y VI De Vivienda: Resultados Definitivos*, 941; Iden, *Estado de la Población Peruana, 2014*, 7; Iden, Mapa de Pobreza, http://censos.inei.gob.pe/Censos2007/Pobreza/.

21   Natteri, Oscar. "EMAPA insiste en donacion de terreno para planta," in *Correo del Sur* January 21, 2015. Accessed 1/21/15: http://icu/diariocorreo.pe/ciudad/emapa-insiste-en-donacion-de-terreno-para-planta-558976/.

22   Thoms, Robins, Ecos, Brooks and Espinoza Gonzales, *Results of June–July, 2015 Field Study*, 7–8, 14.

23   Thoms, Robins, Ecos, Brooks and Espinoza Gonzales, *Results of June–July, 2015 Field Study*, 6, 11–12; Thoms and Robins, 7; Thoms, Robins, Ecos, Brooks and Espinoza Gonzales, *Results of June–July, 2015 Field Study*, 6, 11–12; Thoms, Robins, Ecos and Brooks. *Results of June/July 2016 Assessment of Soil and Fish*, 9–10. See Tables 2, 3, 6 and 8 in appendix. The results are based on three data sets of inductively coupled plasma mass spectrometry (ICPMS) results. One is of fifteen soil/rock samples gathered in different locations in the city of Huancavelica in June/July 2015. Stone samples were gathered by tapping fragments from travertine, while soil was excavated from a depth of approximately one inch. Arsenic concentrations, measured in miligrams per kilogram (mg/kg) had a range of 12 to 1,060, with travertine concentrations ranging from 530 to 887 and cinnabar tailings/soil ranging from 436 to 562 mg/kg. The second data set is of five soil samples collected from a depth of approximately two inches in different places in the city in 2009. The arsenic concentrations were as follows, measured in mg/kg: non detect, 29, 1517, 1233 and 1225. Arsenic concentration in thirty-five adobe samples collected from the interior of ten homes in

heavy metal and cinnabar are often found together in mineral deposits. Consequently, some of the arsenic lurking in the city's soils and earthen homes may result from historic refining activity. During the colonial and national periods, arsenic was so concentrated in some parts of the mine that the cinnabar which it impregnated was left unmined. Although refiners also avoided refining ores with discernable arsenic, given the frequent colocation of cinnabar and arsenic, the latter was inevitably smelted along with cinnabar. Beyond being a carcinogen, arsenic, like mercury, may result in stillbirths, birth deformities, kidney problems, cognitive development issues, and is transmitted in mother's milk.[24]

Similarly, Huancavelica's soils, and some earthen homes, are also contaminated with lead.[25] Like mercury, lead poisoning has a disproportionate effect on the central nervous system and kidneys, and may result in spontaneous abortions, stunted physical and cognitive development, and cancer.[26] Not only has lead been found in the city's soils, but also in its water supply, although the data is conflicting. In one study of four water samples used for human consumption, three tested above the Peruvian screening level of .05 mg/l.[27]

---

different parts of the city in 2010 ranged from 21 to 582, with an average of 175. The US EPA Regional Screening Level for arsenic in residential soil is 0.39 mg/kg for a thirty year period of exposure, however background concentrations globally are often an order of magnitude higher.

24   Hawley, 7; Dueñas, 161; Lohmann Villena, *Las minas de Huancavelica*, 183; Patiño Paúl Ortíz, 349; Contreras and Díaz, 16; United States Environmental Protection Agency, "Arsenic, inorganic," 3–17; http://www.atsdr.cdc.gov/toxfaqs/index.asp.

25   Thoms, Robins, Ecos, Brooks and Espinoza Gonzales, *Results of June–July, 2015 Field Study*, 12. See Tables 3, 6 and 8 in appendix. Lead concentrations in the 2015 dataset of soil and travertine ranged from 7 to 179,300 mg/kg. Calcine materials, including residential floors that appeared to contain calcine, had the highest concentrations ranging from 828 to 179,300 mg/kg. The lowest concentration is twice the US EPA Regional Screening Level of 400 mg/kg per day. In a data set of soil samples collected in 2009 in the city of Huancavelica at a depth of approximately 2 inches and analyzed by ICPMS, the concentrations of lead, measured in mg/kg were 22, 125, 1,802, 2,319, and 15,397. In thirty five adobe samples collected from the interior of ten homes in the city in 2010, the concentrations ranged from 40 to 736, with an average of 277.

26   United States Environmental Protection Agency, "Lead and compounds," 2, 7–9; http://www.who.int/mediacentre/factsheets/fs379/en/.

27   The results were as follows: Sample from spring in Conayca, Huancavelica, contained lead in the amount of .08 mg/l; tap water drawn from Rio Cachi, Izcuchaca, Huancavelica, .110 mg/l; and two samples of tap water drawn from the Rio Ichu, in the city of Huancavelica, which contained lead in the amounts .112 mg/l and .011 mg/l respectively. The screenling level for lead in drinking water established by Peru's Institute of Technological

A second study conducted in 2010 found levels of lead, arsenic, cadmium and copper to be below screening levels in drinking water in all but two samples.[28] Preliminary results of samples collected in 2015 of drinking water in Huancavelica did not contain detectable levels of mercury. In addition, preliminary analysis of a limited number of animal and vegetable foodstuffs from the region, also gathered in 2015, were largely free of mercury and as a result do not appear to pose an imminent intoxication risk to the population. Likewise, trout from the fish farms in Palca, Acoria and the Pultocc Lagoon in 2016 did not contain detectable levels of mercury.[29]

Industrial Research and Technical Standards (Instituto de Investigación Tecnológica Industrial y Normas Técnicas (ITINTEC)) is .05 mg/l, while that of the US EPA is .015 mg/l. See http://water.epa.gov/lawsregs/rulesregs/sdwa/lcr/lcrmr_index.cfm.

28   Máximo Enrique Ecos Lima, "Informe de monitoreo agua de consumo humano 'Distritos de Huancavelica, Izcuchaca y Palca'" (Lima: Companía de Minas Buenaventura, 2010), 3, 7–8. Those that exceded it were drawn from drinking water in the district of Izcuchaca, and contained arsenic at .0176 and .0166 mg/l respectively, above the Peruvian, WHO and US EPA screening level of .01 mg/l . The standards of the Peruvian Government (DS No. 031-2010- SA), World Health Organization and US EPA are .01 mg/ml. See http://water.epa .gov/lawsregs/rulesregs/sdwa/arsenic/regulations.cfm; http://www.who.int/water_sanitation_health/dwq/arsenicsum.pdf.

29   Thoms, Robins, Ecos, Brooks and Espinoza Gonzales, *Results of June-July, 2015 Field Study*, 6–7, 10, 12–13, 17–18. See Tables 4 and 7 in appendix. Five water samples were collected: four from urban drinking water sources and one from the Rio Disparate which is used for clothes washing. Samples were placed into containers provided by CETOX laboratory in Lima, Peru, and kept in a refrigerator or on ice prior to analysis for total mercury. Analysis was by cold vapor atomic adsorption spectroscopy, adhering to method number APHA-AWWA-WEF 3112b, 22nd edition, and the detection limit was .8 µg/l. Food samples were analyzed for total mercury at CETOX laboratory in Lima using cold vapor atomic adsorption spectroscopy, following method number APHA-AWWA-WEF 3112b, 22nd edition. The samples consisted of fifty-two trout samples: thirteen from Acoria's municipal fish farm, twelve each from Palca's municipal fish farm and Sacsamarca fish farm, twelve from Pultocc Lagoon fish farm, three line caught from Choclocoha Lagoon. Additional samples included two of alpaca and one each of beef, lamb, barley, potato and mashua. Fish samples were derived from fillets which were prepared on a clean cutting board with a clean knife, both of which were sanitized between samples. Individual samples were then rinsed with tap water, placed in a sealable plastic bags, and kept refrigerated or on ice until analysis. Mercury was detected only in the trout. One such sample from Lake Chuqlluchucha contained mercury at concentrations of .038 and .046 mg/kg, falling just short of a screening level for a subsistence based diet. The limited presence of mercury in two trout samples drawn from this lagoon suggests that a degree of mercury methylation is taking place in the lagoon. Total mercury is often used as an indicator of methylmercury in fish as most of the mercury in fish tissue is in the methylated form. Mercury was not detected in trout

The levels of mercury, lead and arsenic in the soils and building materials of Huancavelica, and the detection of lead in drinking water, do, however, call for more research. Among the questions is the degree to which chronic exposure to different species of mercury affects people's health. Comparative studies of incidences of kidney problems, birth deformities, mental retardation and neurological disorders may be indicative of mercury's continuing impact on life in Huancavelica. The locally produced food chain also requires more research.

Beyond research, there is a need for a wider recognition of the role that Huancavelica has played in global economic and technical development. Remarkably, the mercury producing sites of Almadén in Spain and Idrija in Slovenia are UNESCO world heritage sites, yet Huancavelica's efforts for such recognition have thus far been unsuccessful.[30] While the designation of Almadén as a Heritage Site recognizes its role in colonial American silver production, this has not been applied to Huancavelica, which was equally important. It is apparent that the Santa Bárbara mines meet several of the criteria for such a designation, although only one is necessary. For example, that Lope de Saavedra's busconil smelter was adopted as the Bustamante smelter in Almadén, and used there until the twentieth century, demonstrates an "important interchange of human values...on developments in...technology." Ironically, it was such smelters that allowed Almadén to achieve the production levels, and ensuing economic and technical importance, which the UNESCO designation recognizes.[31]

As the hemisphere's largest and only legal source of cinnabar extraction in the colonial Americas which powered the rise of global trade networks, the Santa Bárbara site, with its ruins of buildings and smelters, represents a "an outstanding example of [a] technological ensemble...which illustrates (a) significant stage...in human history." Similarly, it is "an outstanding example of a... land-use...which is representative of...human interaction with the environment."[32] The designation of Huancavelica as a UNESCO World Heritage Site would broaden awareness of the city's contribution to global history and economic development as well as the social and environmental legacy with the population contends. It would also promote the region as a touristic destination, which would stimulate economic development.

---

tissue drawn downstream of the city of Huancavelica in Acoria and Palca, nor in other samples above the detection limit of .03 mg/kg.

30 "Heritage of Mercury. Almadén and Idrija," http://whc.unesco.org/en/list/1313.
31 http://whc.unesco.org/en/criteria/.
32 http://whc.unesco.org/en/criteria/.

Beyond research and recognition, there is an urgent need for action to reduce or eliminate the heavy metals to which the residents in the city of Huancavelica are constantly exposed. Home replacement, or at the least, remediation through encapsulation or removal of contaminated soil and adobe, is urgently needed for the estimated 19,000 people who live in such toxic homes. In 2015, a pilot project remediated six mercury contaminated homes in Huancavelica through the application of one and one quarter centimeters thick gesso stucco on the walls and the pouring of a eight centimeter thick cement floor. Testing indicated that, on the whole, mercury vapor levels declined, and indoor sources of contaminated dirt and dust were largely eliminated.[33] Increased paving of streets would also reduce the amount of contaminated dust that people ingest inside and outside of their homes.

If the people and city of Huancavelica are to prosper and enjoy the fruits of health, then large scale investments in infrastructure and social housing will be necessary. Such initiatives would not only foster better public health, but also employment and economic development. Rather than ignoring, minimizing or denying the toxicity with which the city is burdened, Huancavelica's leaders have a responsibility to recognize and use it at a national level to educate policymakers and leverage the resources they need, and deserve, to address these issues. Huancavelica's cinnabar, which the Spaniards saw as a divine blessing, turned out for many to be a mortal curse. Banishing that curse from the homes, land, sediments, air and people of the city could turn out to be the city's greatest economic and public health blessing. Today, however, Huancavelica is being largely bypassed by the global prosperity which it enabled. It remains isolated, neglected, povertous and largely ignored at a national level; wrongly and unnecessarily condemned to the toxic residues of history. Although he was referring to corruption, Viceroy Lemos' words ring true today: "there is much to remedy in Huancavelica."[34]

---

33   Thoms, Robins, Ecos, Brooks and Espinoza Gonzales, *Results of June-July, 2015 Field Study*, 9–10, 16. See Table 5 in appendix.

34   Fernández de Castro y Andrade, "Advertencias que hace el Conde de Lemos a la relación del estado," 255. "Mucho hay que remediar en Huancavelica."

Bibliography

Archival Collections

Archivo-Biblioteca Arquidiocesano "Monseñor Taborga" (ABAS), Sucre, Bolivia
    Archivo Arzobispal, Clero, Tribunal Eclesiástico
Archivo Departmental de Cuzco (ADC), Cuzco, Peru
    Fondo Corregimiento
Archivo General de Indias. (AGI) Seville, Spain
    Indiferente
    Lima
Archivo y Bibliotecas Nacionales de Bolivia (ABNB), Sucre, Bolivia
    Audiencia de La Plata, Minas (ALP Minas)
    Colección Ruck (Ruck)
    Correspondencia, Audiencia de Charcas (CACh)
    Reales Cedulas (RC)
    Sublevación General de Indios (SGI)
Casa Nacional de Moneda Archivo Histórico (CNMAH), Potosí, Bolivia
    Cabildo, Gobierno e Intendencia (CGI)

Archival Primary Sources

Alcalá y Amurrio, Juan. Directorio del Beneficio del Azogue, en los Metales de Plata. Documentos que se dan en sus Reglas. Oruro, 1781, ABNB, Ruck 80.

Auto acordado por la Real Audiencia, sobre que los Corregidores no repartan géneros. La Plata, February 13, 1772. ABNB, EC.1772.59.

Autos criminales seguidos por Lúcas Alanza, indio, contra el Gobernador de Sacaca don Eduardo Ayaviri, por haberle inferido una paliza que casi le compromete su vida y varias injurias, además dicho Gobernador dispone mal de los Reales Tributos. La Plata, June 10, 1758. ABNB, EC.1758.22.

Autos seg.s por los indios del Pueblo de Guagui contra el cura Dr. Don Pedro Márquez sobre varios ynjurias que les infirio. La Plata, May 18, 1754. ABNB, EC.1754.49

Autos seguidos por los Curas de este Arzobispdo y demas sufraganeos, sre la suspencion de la Rl. Cedula que manda que los Ynds no paguen obvenciones. La Plata, September 25, 1760. ABNB, EC.1760.75, 29.

Avisos muy importantes y noticias muy particulares que de un bien intencionado y deseoso del mayor revicio... sobre el mayor regimen, estableciemiento y gobierno de la real mita. Potosí, January 20, 1762. ABNB, ALP Minas 151/10.

Capítulos de una carta escrita por esta Real Audiencia a su majestad: De su parecer sobre si convendría traer negros a Potosí para aliviar a los indios de la mita y hacer que algunos aspectos de ésta sean resueltos exclusivamente por el alcalde mayor de minas sin intervención del corregidor de dicha villa. La Plata, February 1, 1610. ABNB, ALP Minas 123/4.

Capítulos puestos por los indios del pueblo de Calacoto, provincia de Pacajes, contra su cacique don Juan Machaca, y José Rivera, escribano publico de dicha provincia, sobre exacciones, malos tratamientos con motivo de la mita de Potosí, defraudaciones de tributes, etc. 1747–1750. ABNB, ALP Minas 126.20.

Capítulos puestos por Pedro Pirua y otros indios del pueblo de Chayanta, provincia del mismo nombre, contra su gobernador don Sebastián Auca, sobre defraudación de mitayos y otros excesos. Potosí, 1757–1758. ABNB, ALP Minas 127/14.

Carta de Juan Gutiérrez de León y Matías Díaz Rodo, oficiales reales en el asiento del Espíritu Santo, provincia de Carangas, a esta Real Audiencia: Los excesos cometidos por don Nicolás de Avalos y Ribera, corregidor de dicha provincia, contra Sebastían Cabezudo de Velasco, el principal ingeniero y aviador de aquélla, y contra los indios han hecho descaecer las labores de minas y disminuir los reales quintos. Carangas, September 16, 1664. ABNB, ALP Minas 96/5.

Carta de Pedro Camargo. Lima, March 12, 1595. AGI, Lima, 35, #2.

Carta de su presidente, el licenciado Alonso Maldonado de Torres, asistente en Potosí, a esta Real Audiencia: Recomienda el mantener los indios de mita asignados al trajín de plata y azogue entre Potosí y Arica. Potosí, November 16, 1606. ABNB, ALP Minas 123/1.

Carta del Cura Vicario del Tomavi, provincia del Porco, a la Rl. Aud de la Plata. Tomavi, November 13, 1616. ABNB, CACh 728.

Carta del Virrey del Perú # 38, "Da cuenta a V.Md. de todo lo que en materia de azogue sea escrito." February 2, 1630, AGI, Indiferente general 1777.

Carta del Virrey Velasco al Rey, Lima, May 5, 1600, AGI Lima 34.

Damián de Jeria, Protector de los Naturales. Lima, January 10, 1604. AGI Lima, 34, No. 42.C.

Don Francisco José Ayra de Ariuto, cacique del pueblo de Pocoata, parcialidad de Hurinsaya, provincia de Chayanta, sobre los excesos del teniente de gobernador de los minerales de Titiri y Aullagas para con los indios. La Plata, 1691. ABNB, ALP Minas 147.4

Don José de la Rúa, protector de naturales de Potosí, sobre el indebido cobro de 52 pesos anuales que el subdelegado del partido de Carangas, pretende llevar con el título abusivo de rezagos de mita, a cada indio de los que, estando repartidos para dicho servicio, no asistieron a él en virtud del indulto concedido últimamente por las cortes extraordinarias. Potosí, November 28, 1814. ABNB, ALP Minas 130.10.

Don Juan de Dios Cavitas, indio principal del pueblo de Turco, provincia de Carangas, pidiendo se rebaje la contribución de mitayos del ayllo Ilanaca, parcialidad de Pumiri, en dicho pueblo, por falta de indios aptos. April 16, 1807, ABNB ALP Minas 130/8.

Don José Salas Ordóñez, visitador de la provincia de Sicasica, contra el cacique Felipe Alvarez. La Plata, September 8, 1765. ABNB, EC.1765.77.

Don Rodrigo de Mendoza y Manrique, administrador y arrendatario que fue de las minas y los ingenios de don Pedro Sorez de Ulloa en el cerro y la ribera de Potosí, con el maestro de campo don Rodrigo Campuzano, hermano de doña Francisca Campuzano, viuda y heredera que fue de dicho don Pedro, sobre la liquidación de los pesos impendidos en el avío de las haciendas mencionadas durante el tiempo de su administración y arrendamiento. La Plata, February 5, 1656–1669. ABNB, ALP Minas 15/1.

Don Ventura de Santelices y Venero, corregidor de Potosí, sobre la falla de 54 mitayos que anualmente acusa la provincia de Cochabamba con grave perjuicio para la real hacienda. Cochabamba, September 1, 1755 – La Plata, February 27, 1756. ABNB, ALP Minas 127/10.

Emetherio Ramírez de Arellano. Huancavelica, April 3, 1649. AGI, Lima, 279–4.

Expediente de las Diligencias practicadas en virtud de Real Provision por el Lisenciado Dn Custaquio Ferrera contra el Dor Dn Joseph de Barco y Oliva Cura del Beneficio de Cicacica, sobre la exaccion excesiva de Derechos Parrochiales y otros abusos a los indios. La Plata, April 20, 1769. ABNB, EC.1769.58.

Expediente instruído con las representaciónes de varios indios mitayos del pueblo de Capinota, provincia de Cochabamba, sobre los tributos que indebidamente se les exigen por los años que sirvieron la mita de Potosí, October 27, 1792. ABNB, ALP Minas 129/2.

Expediente promovido por don Nicolás de Sarabia y Mollinedo asentista de la Real Mina de azogue de Guancavelica sre que los corregidores y gobernadores de las provincias que tienen obligación de remitir mitas de indios para el trabajo de otra mina lo ejecuten sin dilación alguna y sre lo ocurrido con este motivo en la de Jauja a causa de haberse intentado cumplir con la remisión de mitayos aquel gobierno como se le mandó que avisa la resistencia que hicieron e insultos al cobrador enterador de mitas don Jacinto Maita. Lima, November 10, 1811. AGI, Indiferente 1335.

Expediente seguido por don Roque de Reinalte, sobre la prisión a que le han reducido por cierta deuda no obstante de su calidad de minero en Potosí. La Plata, August 14–19, 1697. ABNB, ALP Minas 19/10.

Expediente sobre la postura para la conducción de azogue de esta Caja de Potosí a las demas del Virreinato. Potosí, 1781. CNMAH, CGI/M-64/26.

Felipe Quispe, Isidro Quispe, Andrés Flores, Lucas Choque y Baltasar Aruquipa, indios del publo de Guarina, provincia de Omasuyos, sobre los avíos que su cacique omitio darles para venire a la mita de Potosí. La Plata, 1750–1775. ABNB, ALP Minas 128/2.

Gaspar de Carvajal, Alonso de la Cerda, y Miguel Adrián al Rey, March 17, 1575, AGI, Lima, 314.

Informe del corregidor don Juan Medrano Navarrete sobre la situación de los pueblos que componen la provincia de Pacajes. In *Potosi: La version aymara de un mito europeo. La mineria y sus efectos en las sociedades andinas del siglo XVII (La Provincia de Pacajes)*. Teresa Cañedo-Arguelles Fábrega. Madrid: Editorial Catriel. 1993, 109–111.

Lorenzo Mateo, indio del pueblo de San Juan de Challapata, ayllo Jilha, provincia de Paria, pidiendo se le exima de la mita de Potosí por estar enfermo y baldado. La Plata, February 23, 1736. ABNB, ALP Minas 126/16.

Los indios de las parcialidades de Laymes, Cullpas, Chayantacas, Carachas, Puracas y Sicoyas, de la provincia de Chayanta, sobre extorciones a contribuyentes de la mita y otros excesos de los caciques. La Plata, 1797–1799. ABNB, ALP Minas 129/16.

Marqués de Casa Concha. Relación del estado que ha tenido y tiene la real mina de Guancavelica. Lima, 1726, AGI, Lima 469.

Memorial del capitán Don Pedro Gutíerrez Calderón de algunas advertencias considerables al servicio de Dios Nuestro Señor y de su Real Magestad, Lima, May 2, 1623, AGI, Lima, 154.

Memorial que el licenciado Gaspar González Pavón escribió, de Potosí a España, a don Gómez Dávila, corregidor provisto de dicha Villa: Refiérle principalmente al régimen de la mita. Potosí, January 25, 1658. ABNB, ALP Minas 125/14.

Miguel Flores, Luis Flores y Santos García, indios del pueblo de Chayanta, provincia del mismo nombre sobre el exceso con que se trate de enviarlos a la mita de Potosí antes de los dos años de descanso. La Plata, September 1, 1757. ALP, Minas 127/12.

No Title, La Plata, January 11, 1776. ABAS, Archivo Arzobispal, Clero, Tribunal Eclesiástico.

Provisión circular expedida por esta Real Audiencia pra que los gobernadores de indios cesen en la costumbre de substituir a los contribuyentes acomodados con otros infelices en el servicio de la mita de Potosí. La Plata, 1773. ABNB, ALP Minas 127/22.

Provisión de don Melchor de Navarra y Rocaful, Duque de la Palata virrey del Perú: En cumplimiento de la Real Cédula de 1676.12.08, se extiende la obligación de dar indios mitayos para las minas e ingenios de Potosí a los pueblos que hasta ahora estaban exentos y se señala el orden general que ene ése y otros puntos ha de tener este servicio. Lima, December 2, 1688. ABNB Ruck 1/4:446–453.

Provisión del virrey del Perú don Melchor Portocarrero Lazo de la Vega, conde de la Monclova: Establece un nuevo régimen para la mita de Potosí, abandonando – por los numerosos inconvenientes que sobrevinieron en su aplicación – el régimen que había establecido su antecessor, don Melchor de Navarra y Rocafull, duque de la Palata. Lima, April 27, 1692. ABNB, Ruck 11, 23–37.

Provisión del virrey Diego Fernández de Córdoba, sobre el hospital real de la Villa de Huancavelica. Cuzco, March 21, 1623. ADC, Fondo Corregimiento, Legajo No. 7, No. 114, Cuaderno 3.

Real Cedula. Aranjuez, May 2, 1752. ABNB, RC.571.

Real Cédula a los Oficiales Reales de Potosí: Se ha mandado al Virrey del Perú que haga corer la mita solamente en las 16 provincias primativamente afectadas y ya no en las que después se agregaron con otras disposiciones, sobre la proporción exigible a los pueblos, pago de leguajes y jornales, asistencia de un oidor en Potosí encargado de la defensa de los mitayos, etc. Seville, October 22, 1732. ABNB, ALP RC 545.

Registro de entrega de los indios de mita de las provincias de Porco, Canas y Canches, Chucuito, Chayanta, Paria, Asángaro y Umasuyo, por el capitán general de este servicio, don Juan José de Orense, a los dueños de minas e ingenios a quienes corresponden. Potosí, June 24–November 1, 1736. ABNB, Ruck 11, 1–22.

Representación de Dn. Mathias Chuquimanqui, Gobernador I Casique de Caquiviri, provincia de Pacajes, contra los españoles que en sus haciendas, hacen pasar indios sin pago de la tasa ni concurrencia a la mita. La Plata, September 28, 1743, ABNB, ALP Minas 149/14.

Representacion de la ciudad del Cusco, en el año de 1768, sobre excesos de corregidores y curas. In Colección Documental de la Independencia del Perú. Vol. I, Book II. "La Rebelión de Túpac Amaru: Antecedentes." Cárlos Daniel Válcarcel, Ed. Lima: Comisión Nacional del Sesquicentenario de la Independencia del Perú, 1971.

Simón Pérez, indio forastero en el pueblo de Asangaro, provincia del mismo nombre, sobre que se le asignen tierras de repartimiento en dicho puebo, como a mitayo de él. Asángaro, August 23, 1752. ABNB, ALP Minas 127/2.

Testimonio de la representacion que hizo el senor Fiscal Promotor General por parte de don Lorenzo Apu Bedoya y Melchora de la Cruz Anaya, indios del pueblo de Toledo sobre visitadores eclesiásticos. December 16, 1753, La Plata. ABNB, EC.1753.144.

Testimonio de los informes que a instancia del doctor don Victorián de Villava, fiscal de esta Real Audiencia, expidieron don Francisco de Viedma, gobernador intendente de Puno; el doctor Felipe Antonio Martínez de Iriarte, cura propio de la doctrina de Chaqui, partido de Porco, y vicario pedáneo de Potosí; y el doctor don José de Osa y Palacios, cura propio que fue de la doctrina de Moscarí, partido de Chayanta, sobre los perjuicios que a los pueblos de indios de dicha circunscripción se siguen de la mita de Potosi. La Plata, November 24, 1794. ABNB, ALP. Minas 129.8.

Testimonio de los obrados relativos a los mitayos que deben mandar a Potosí, para el trabajo de minas del lugar de Sicasica. Potosí, 1759. CNMAH, CGI/M-65/17.

Testimonio en fi6 de las cartas de los rebeldes, comisiones e informe que Diego Cristóbal Túpac-Amaru hizo al Exmo. Sr. Virrey de Lima, en respuesta del indulto Gral que libro. Peñas, November 15, 1781. ABNB, SGI.1781.248.

Vista que el licenciado don Martín de Arriola, oidor de la Real Audiencia, tomó del ingenio nombrado Nuestra Señora de Guadalupe, provincia de los Chichas, propio del Capitán Pedro de Espinoza y Ludueña para establecer las condiciones del trabajo. June 20, 1630. ABNB, ALP Minas 131/2.

Printed Primary Sources

Acosta, José de. *De procuranda indorum salute.* L. Pereña et al., Eds. Madrid: Consejo Superior de Investigaciones Científicas, 1984.

Acosta, José de. *Historia natural y moral de las Indias.* José Alcina Franch, Ed. Madrid: Historia 16, 1987.

Agia Miguel, Fray, *Servidumbres personales de indios.* Javier de Ayla, Ed. Sevilla: Escuela de Estudios Hispanos-Americanos, 1946.

Agricola, Georgius *De Re Metalica.* Herbert C. and Lou H. Hoover, Eds and Trans. New York: Dover Publications, 1950.

Álvarez de Toledo y Leyva, Marqués de Mancera, Pedro. "Relación del estado del gobierno del Perú que hace el Marqués de Mancera al Señor Virrey Conde de Salvatierra." In *Colección de las memorias o relaciones que escribieron los virreys del Perú.* Vol. II. Ricardo Beltrán y Rózpide, Ed. Madrid: Imprenta del Asilo de Huérfanos del S.C. de Jesús, 1921, 125–209.

Arias de Ugarte, "Carta a S.M. del nuevo oidor doctor Arias de Ugarte dando cuenta del estado en que halló la Audiencia de Charcas. Acompaña un memorial de los indios de su distrito en razón de los agravios que reciben. Pide que se tomen medidas para que no desembarquen extrajeros y gente perdida por el puerto de Buenos Ayres." Potosi, February 28, 1599. In *La audiencia de Charcas: correspondencia de presidentes y oidores. Documents del Archivo de Indias.* Vol. III. Roberto Levillier, Ed. Madrid: Imprenta de Juan Pueyo, 1922, 355–367.

Armendaris, Marques de Castel-Fuerte, José. "Relacion del estado de los reynos del Perú que hace el Excmo. Señor Don José Armendaris, Marqués de Castel-Fuerte, á su successor el Marqués de Villagarcía, en el año de 1736." In *Memorias de los virreyes que han gobernado el Perú durante el tiempo del coloniaje español.* Vol. III. M.A. Fuentes, Ed. Lima: Librería Central de Felipe Bailly, 1859, 1-3d69.

Arzáns de Orsúa y Vela, Bartolomé. *Historia de la Villa Imperial de Potosí.* 3 Vols. Lewis Hanke and Gunnar Mendoza, Eds. Providence, RI: Brown University Press, 1965.

Audiencia de Lima. "Relacion que la Real Audiencia de Lima hace al excelentísimo Sr. Marqués de Castel-Dosrius, Virey de estos reinos, del estado de ellos, y tiempo que ha gobernado en vacante." In *Relaciones de los Virreyes y Audiencias que han Gobernado el Perú.* Vol. II. Sebastian Lorente, Ed. Madrid: Imprenta y Estereotipia de M. Rivadeneyra, 1871, 283–307.

Avendaño, Diego de. *Thesaurus Indicus*. Angel Muñoz García, Trans. and Ed. Pamplona: Ediciones Universidad de Navarra, 2001.

Ayans, Antonio de. "Breve relación de los agravios que reciven los indios que ay desde cerca del Cuzco hasta Potosí, que es lo major y más rico del Perú, hecha por personas de muchas experiencia y Buena conciencia y desapasionadas de todo interés temporal y que solamente desean no sea Dios N.S. tan ofendido con tantos daños como los indios resciven en sus almas y haziendas y que la conciencia de Su Magestad se descargue mejor y sus Reales Rentas no sean defraudadas en nada sino que antes bayan siempre en continuación (1596)." In *Pareceres juridicos en asuntos de indias (1601–1718)*. Ruben Vargas Ugarte, Ed. Lima: CIP, 1951, 35–88.

Baquijano, Joseph. "Historia del descubrimiento del cerro de Potosí, fundación de su Imperial Villa, sus progresos y actual estado." In *Mercurio Peruano*, 7 (1793): 25–48.

Blanco, José María. *Diario de viaje del Presidente Orbegoso al sur del Perú*. 2 Vols. Felix Denegri Luna, Ed. Lima: PUCP-IRA 1974.

Borja, Príncipe de Esquilache, Francisco de. "Relación que hace el Príncipe de Esquilache al Señor Marqués de Guadalcasar, sobre el estado en que deja las provincias del Perú." In *Memorias de los virreyes que han gobernado el Perú durante el tiempo del coloniaje español*. Vol. I. M.A. Fuentes, Ed. Lima: Librería Central de Felipe Bailly, 1859, 71–145.

Brown, Lester Jr,. "Peru." In *Minerals Yearbook Area Reports: International, 1967*. Vol. 4. Washington, DC: Bureau of Mines, 1969, 601–614.

Burgess, John, Sumner Anderson and R. Lester Jr. "Peru." In *Minerals Yearbook Area Reports: International 1963*. Vol. IV. Washington, DC: Bureau of Mines, 1964, 303–324.

Calancha, Antonio de la. *Crónica moralizada del orden de San Augustín en el Perú, consucesos egemplares en esta monarquía*. Lima: Universidad Nacional Mayor de San Marcos, 1974.

Cañete y Domínguez, Pedro Vicente. *Guía histórica, geográfica, física, política, civil y legal del gobierno e intendencia de la provincia de Potosí*. Potosí, Bolivia: Editorial Potosi, 1952.

Cantos de Andrade, Rodrigo. "Relación de la Villa Rica de Oropesa y minas de Guancavelica." In *Relaciones geográficas de Indias*. Vol. I. Marcos Jiménez de Espada, Ed. Madrid: Ediciones Atlas, 1965, 303–309.

Capoche, Luis. *Relación general de la Villa Imperial de Potosí*. Lewis Hanke, Ed. Madrid: Ediciones Atlas, 1959.

Caravantes, Francisco López. *Noticia general del Perú*. 6 vols. Marie Helmer, Ed. Madrid: Ediciones Atlas, 1989.

Carranza, L. "De Huanta á Lima por el camino de Huancavelica – Año de 1866." In *Boletín de la Sociedad Geográfica de Lima*. No. 5 (September 30, 1895): 176–187.

Carrió de la Vandera, Alonso. *El lazarillo de ciegos caminantes*. Caracas: Biblioteca Ayacucho, 1985.

Carrió de la Vandera, Alonso. "Plan de Gobierno del Perú." In *El lazarillo de ciegos caminantes*. Caracas: Biblioteca Ayacucho, 1985, 231–292.

"Carta de fray Domingo de Santo Tomás al Consejo de Indias." In *Fr. Domingo de Santo Tomás: Defensor y Apostol de los Indios del Perú. Su vida y sus escritos*. José María Vargas, O.P., Ed. Quito: Editorial Santo Domingo, 1937, 74–84.

Charles II. "Confirmación de Su Magestad del Asiento que hizo el Excelentísimo Señor Duque de la Palata con los mineros de Guancavelica sobre la labor, y beneficio de la mina de azogue con las condiciones y calidades que se refieren." Madrid, June 10, 1685. In *Minas e indios del Perú, siglos XVI-XVIII*. Nadia Carnero Albarrán, Ed. Lima: Universidad de San Marcos, 1981, 157–200.

Cieza de León, Pedro. *Crónica del Perú*. Lima: Pontífica Universidad Católica del Perú, 1984.

Cobo, Bernabé. *Historia del Nuevo Mundo*. Francisco Mateos, S.J., Ed. Madrid: Ediciones Atlas, 1956.

Contestación al discurso sobre la mita de Potosí escrito en La Plata 9 de marzo de 1793 contra el servicio de ella. In María del Carmen Cortés Salinas, "Una polémica en torno a la mita de Potosí a fines del siglo XVIII," in *Revista de Indias*. Vol. XXX, Nos. 119–122 (January–December, 1970), 138–175.

Cosme Bueno y Alegre, Francisco Antonio. Cosme. *Geografía del Perú Virreinal (Siglo XVIII)*. Carlos Daniel Valcárcel, Ed. Lima: NP, 1951.

Crosnier, León. "Geologie du Perou. Notice geologique sur les Departments de Huancavelica et d'Ayachcho." In *Annales de mines*. 5th series, Vol. II (1852): 1–107.

Cueva, Conde de Castellar, Baltasar de la, "Relacion general que el Excmo. Señor Conde de Castellar, Marqués de Malagon, Gentilhombre de la Cámara de su Majestad, de su Consejo, Cámara y Junta de Guerra de Indias, Virey, Gobernador y Capitan General que fué de estos Reinos, hace del tiempo que los gobernó, estado en que los dejó, y lo obrado en las materias principales con toda distinction." In *Memorias de los virreyes que han gobernado el Perú durante el tiempo del coloniaje español*. Vol. I. M.A. Fuentes, Ed. Lima: Librería Central de Felipe Bailly, 1859, 147–259.

Dueñas, Enrique, "Fisionomia minera de las provincias de Tayacaja, Angaraes y Huancavelica." *Boletín del cuerpo de ingenieros de minas del Peru*. No. 62. Lima: El Lucero, 1908.

Escovedo, Jorge. *Proyecto, que sobre la extinción de repartos, y modo de verificar losPiadosos Socorros, que la generosa bondad del Rey nuestro señor quiere se franqueen a los indios*. Lima: Imprenta Real, 1784.

Fernández de Cabrera, Conde de Chinchón, Luis Jerónimo. "Relación del estado en que el Conde de Chinchón deja el gobierno del Perú al Señor Virrey Marqués de Mancera." In *Colección de las memorias o relaciones que escribieron los virreys del Perú*. Vol. II. Ricardo Beltrán y Rózpide, Ed. Madrid: Imprenta del Asilo de Huérfanos del S.C. de Jesús, 1921, 46–124.

Fernández de Castro y Andrade, Conde de Lemos, Pedro. "Advertencias que hace el Conde de Lemos a la relación del estado del reino que le entregó la Real Audiencia de Lima del tiempo que gobernó en vacante de virrey que fue de año y más de ocho meses, dirigida a la reina nuestra señora en el real y supremo consejo de Indias." In *Los virreyes españoles en America durante el gobierno de la casa de Austria. Perú.* Vol. IV. Lewis Hanke and Celso Rodríguez, Eds. Madrid: IMNASA, 1978, 251–271.

Fernández de Castro y Andrade, Conde de Lemos, Pedro. "Carta del Conde de Lemos a S.M. sobre la Mita de Potosí." In *Pareceres juridicos en asuntos de indias (1601–1718).* Ruben Vargas Ugarte, Ed. Lima: CIP, 1951, 155–165.

Fernández de Castro y Andrade, Conde de Lemos, Pedro. "El Conde de Lemos da cuenta a S.M. del estado en que hallo el reino del Perú cuando entró a gobernarlo y el remedio que ha comenzado a poner en las materias más principales de su Gobierno." Lima, March 3, 1668. In *Los virreyes españoles en America durante el gobierno de la casa de Austria. Perú.* Vol. IV. Lewis Hanke and Celso Rodríguez, Eds. Madrid: IMNASA, 1978, 271–273.

Fernández de Castro y Andrade, Conde de Lemos, Pedro. "Refiere lo obrado en poco más de un áno de servicio." January 20, 1669. In *Los virreyes españoles en America durante el gobierno de la casa deAustria. Perú.* Vol. IV. Lewis Hanke and Celso Rodríguez, Eds. Madrid: IMNASA, 1978, 273–276.

Fernández de Villabos, Gabriel. *Vaticinios de la pérdidia de las Indias y mano de relox.* Caracas: Instituto Panamericano de Geografía e Historia, 1949.

Fonseca, George de. "Informe de George de Fonseca del 24 de Julio de 1622." In *Huancavelica colonial: Apuntes históricos de la ciudad minera más importante del Virreynato Peruano.* Mariano Patiño Paúl Ortíz. Lima: Huancavelica 21, 2001, 347–352.

García de Castro, Lope. "Carta a S.M. del Licenciado Castro acerca de las minas y del trabajo de los indios; la sucesión de encomiendas; guerra contra los indios en Chile; fundación de un monasterio de monjas; y otros asuntos de menor importancia." In *Gobernantes del Perú: Cartas y Papeles siglo XVI.* Vol. III. Roberto Levillier, Ed. Madrid: Sucesores de Rivadeneyra, 1921, 287–296.

Gastelumendi, A G. *Huancavelica Como Región Productora De Mercurio.* Lima: Imp. Torres Aguirre, 1920.

Government of Peru. *Almanaque De Huancavelica.* Lima: Instituto Nacional de Estadística e Informática, 2002.

Government of Peru. *Censo General De La República Del Perú, Formado En 1876.* Vol. IV. Lima: Imp. del Teatro, 1878.

Government of Peru. *Censo Nacional De Población Y Ocupación, 1940.* Vol. VI. Lima: Dirección Nacional de Estadística y Censos, 1949.

Government of Peru. *Censos Nacionales, Vii De Población, Ii De Vivienda, 4 De Junio De 1972.* Vol. I. Lima: Oficina Nacional de Estadística y Censo, 1974.

Government of Peru. *Censos Nacionales 1993, IX De Población, IV De Vivienda. Resultados Definitivos. Departamento d Huancavelica.* No. 19. 2 Vols. Lima: Instituto Nacional de Estadística e Informática, 1994.

Government of Peru. *Censos Nacionales 2007: XI de Poblacion y VI de Vivienda. Directorio Nacional de Centros Poblados. Vol. III.* Lima: Instituto Nacional de Estadística, 2010.

Government of Peru. *Censos Nacionales 2007: XI de Población y VI de Vivienda. Principales Indicadores Demográficos, Sociales Y Económicos a Nivel Provincial Y Distrital Huancavelica.* Lima: Instituto Nacional de Estadística e Información, 2009.

Government of Peru. *Censos Nacionales 2007, XI De Población Y VI De Vivienda: Resultados Definitivos.* Lima: Instituto Nacional de Estadística e Informática, 2008.

Government of Peru, *Censos Nacionales 2007: XI de Población y VI de Vivienda. Sistema de consulta de resultados censales. Cuadros estadistcos.* http://censos.inei.gob.pe/cpv2007/tabulados/#.

Government of Peru. *Centros Poblados: Sexto Censo Nacional De Población, Primer Censo Nacional De Vivienda, 2 De Julio De 1961.* Vol. II. Lima: Dirección Nacional de Estadística y Censos, 1966.

Government of Peru. *Compendio Estadístico Departamental.* Lima: Instituto Nacional de Estadística e Informática, 1998.

Government of Peru. *Encuesta Demográfica Y De Salud Familiar 2000: Huancavelica.* Vol. VIII. Lima: Instituto Nacional de Estadística e Informática, 2001.

Government of Peru. *Estado de la Población Peruana, 2014.* Lima: Instituto Nacional de Estadística e Informática, 2014.

Government of Peru. *Inventario Y Evaluación De Los Recursos Naturales De La Zona Altoandina Del Perú: Reconocimiento, Departamento De Huancavelica.* 2 Vols. Lima: Oficina Nacional de Evaluación de Recursos Nacionales, 1984.

Government of Peru. *Población, Mujer Y Salud. Resultados de la encuesta demográfica y de salud familiar, 1996.* Lima: Instituto Nacional de Estadística e Informática, 1997.

Government of Peru. Sistema de Consulta de Principales Indicadores de Pobreza, Mapa de Pobreza, Instituto Nacional de Estadística, http://censos.inei.gob.pe/Censos2007/Pobreza/.

Guaman Poma de Ayala, Felipe. *Nueva crónica y buen gobierno.* Vol. II. Luis Bustios Galvez, Ed. Lima: Editorial Cultura, 1966.

Habich, Eduardo de, "Industria de azogue." In *Boletín de Minas, Industrias y Construcciones.* Vol. II (Lima, 1885): 11–13.

Haenke, Tadeo. *Descripción del Perú.* Lima: Imprenta El Lucero, 1901.

Jiménez, Carlos P. "Estadistica Minera en 1917." In *Boletín del cuerpo de ingenieros de minas del Peru.* No 95. Lima: Imprenta Americana, 1919.

Juan, Jorge and Antonio de Ulloa. *Discourse and Political Reflections of the Kingdoms of Peru.* Norman: University of Oklahoma Press, 1978.

Lanuza y Soltelo, Eugenio. *Viaje ilustrado a los reinos del Perú*. Lima: Pontífica Universidad Católica del Perú, 1998.

Liñan y Cisneros, Melchor. "Relacion de Don Melchor de Liñan y Cisneros, dada al Señor Duque de la Palata, del tiempo de tres años y cuatro meses que gobernó, desde 1678 hasta 1681." In *Memorias de los virreyes que han gobernado el Perú durante el tiempo del coloniaje español*. Vol. I. M.A. Fuentes, ed. Lima: Librería Central de Felipe Bailly, 1859, 261–379.

Lizárraga, Reginaldo de. *Descripción del Perú, Tucumán, Río de la Plata y Chile*. Buenos Aires: Union Académique Internationale/ Academia Nacional de la Historia, 1999.

Llano Zapata, José Eusebio. *Memorias histórico, fisicas, critico, apologéticas de la América Meridonal*. Ricardo Ramírez Castañeda, et al., Eds. Lima: Instituto Francés de Estudios Andinos, 2005.

López de Velasco, Juan. *Geografía y descripción universal de las Indias*. Madrid: Establecimiento Tipográfico de Fortanet, 1894.

Loveday, Santiago. "El azogue de Huancavelica." In *Boletín de minas, industrias y construcciones*. Vol. IIo (Lima: 1905): 82–84.

Manso de Velasco, Conde de Superunda, José Antonio. "Relacion que escribe el conde de Superunda, Virrey el Perú, de los principales sucesos de su gobierno, de Real Orden de S.M. comuncada por el Excmo. Sr. Marqués de la Ensenada, su secretario del Despacho universal, con fecha 23 de Agosto de 1751, y comprehende los años desde 9 de Julio de 1745 hasta fin del miso mes en el de 1756." In M.A. Fuentes, ed., *Memorias de los virreyes que han gobernado el Peru durante el tiempo del coloniaje español*. Vol. IV. (Lima: Librería Central de Felipe Bailly, 1859), 1–340.

Markham, Clements. *Cuzco: A journey to the Ancient Capital of Peru; with an Account of the History, Language, Literature, and Antiquities of the Incas. And Lima: A Visit to the Capital and Provinces of Modern Peru*. London: Chapman and Hall, 1856.

Martínez y Vela, Bartolomé, *Anales de la Villa Imperial de Potosí*. La Paz: Imprenta Artistica, 1939.

Martino, Orlando. *Minerals Yearbook Area Reports: International 1976*. Vol. III. Washington, DC, Bureau of Mines, 1976, 821–835.

Matienzo, Juan de. "Carta a S.M. del licenciado Matienzo, con noticia de la residencia, que por encargo del Virrey, habia tomado al corregidor, alcaldes, oficiales y otros jueces de la Villa de Potosi. Describe el estado en que hallo las minas y lo que hizo para aumentar las rentas reales. Refiere el casamiento de Juan de Torres de Vera con la hija del adelantado Ortiz de Zarate, y aconseja que para el major gobierno de las Provincias del Tucumán y Paraguay, se junten en una sola, y se funden pueblos en el Tucumán y en el Rio de la Plata para el comercio directo con España." Potosí, December 23, 1577. In *La audiencia de Charcas: correspondencia de presidentes y oidores. Documents del Archivo de* Indias. Vol. I. Roberto Levillier, Ed. Madrid: NP, 1918, 455–465.

Matienzo, Juan de. *Gobierno de Perú*. Guillermo Lohmann Villena, Ed. Lima: Institut Fracncais D'Études Andines, 1967.

Mendoza y Luna, Juan de, Marqués de Montesclaros. "Carta del Virrey Marqués de Montes Claros a S.M. en materia de Real Hacienda, cantidades que se envian de todo género de hacienda; ruina del Cerro de Guancavelica para lo que pide socorro de azogues; estado que tiene el edificio de la iglesia de Lima, tiempo en que podrá terminar y lo que ha costado a S.M." In "Cuatro cartas del Marqués de Montesclaros referentes a la mina de Huancavelica." Manuel Moreyra y Paz Soldán, Ed. *Revista Histórica* 18 (1949): 89–92.

Mendoza y Luna, Juan de, Marqués de Montesclaros. "Carta del Virrey Marqués de Montes Claros a S.M. informando extensamente sobre las minas de Guancavelica, en virtud de la comunicación y conferencias que sobre el asunto tuvo con su antecesor en aquel gobierno D. Luis de Velasco." In "Cuatro cartas del marque´s de Montesclaros referentes a la mina de Huancavelica." Manuel Moreyra y Paz Soldán, ed. *Revista Histórica* 18 (1949): 93–99.

Mendoza y Luna, Juan de, Marqués de Montesclaros. "Relación del estado del gobierno de estos reinos que hace el excmo. Señor Don Juan de Mendoza y Luna, Marqués de Montesclaros, al excmo. Señor Príncipe de Esquilache su sucesor." In *Memorias de los virreyes que han gobernado el Perú durante el tiempo del coloniaje español*. Vol. I. M.A. Fuentes, Ed. Lima: Librería Central de Felipe Bailly, 1859, 1–69.

Mesía Venegas, Alfonso. "Memorial del P. Alfonso Mesía Venegas, sobre la Cédula del servicio personal de los indios. 1603." In *Pareceres juridicos en asuntos de indias (1601–1718)*. Ruben Vargas Ugarte, Ed. Lima: CIP, 1951, 94–115.

Meyer, Helena M. and Gertrude Greenspoon. "Mercury." In *Minerals Yearbook: 1950*. Washington, DC: Bureau of Mines, 1953, 773–786.

Meyer, Helena M. and Gertrude Greenspoon. "Mercury." In *Minerals Yearbook: Metals and Minerals (except fuels) 1953*. Washington, DC: Bureau of Mines, 1956, 769–788.

Montesinos, Fernando. *Anales del Perú*. Victor M. Maurtua, Ed. Vols. 1–2 Madrid: Imprenta de Gabriel L y del Horno, 1906.

Mugica, Martín José de. "Abusos de varias clases de mitas y carácter perezoso del Indio." In "Las mitas de Huamanga y Huancavelica." Luis Basto Girón. In *Perú Indígena*, Vol. V, No. 13 (December, 1954): 223–241.

Muñiz, Pedro. "El Dr. Muñiz de Lima sobre el serujo de los Indios." In "Pedro Muñiz, Dean of Lima and the Indian Labor Question (1603)." In *Hispanic American Historical Review*, K.V. Fox, Vol. IV2, No. 1 (1962): 75–86.

Murua, Martín de. *Historia general del Perú*. Manuel Ballesteros Gaibrois, Ed. Madrid: Historia 16, 1987.

NA. Descripción de la villa y minas de Potosí. Año de 1603. In *Relaciones geograficas de Indias*. Vol. II. Marcos Jiménez de Espada, Ed. Madrid: Ediciones Atlas, 1965, 372–385.

NA. *Descripción del virreinato del Perú: Crónica inedita de comienzos del siglo XVII.* Boleslao Lewin, Ed. Rosario, Argentina: Universidad Nacional del Litoral, 1958.

NA. "Historia de la mina de Huancavelica." In *Mercurio Peruano* (January 30, 1791): 65–68.

N.A. "Informe." Azángaro, October 18, 1781. In *Colección de obras y documentos relativos a la historia antigua y moderna de las provincias de Río de la Plata.* Vol. IV. Pedro de. Angelis, Ed. Buenos Aires: Librería Nacional de J. Lajouane & Cia, 1910.

N.A. Attributed to Mariano Eduardo de Rivero y Ustáriz. "Memoria sobre la mina de azogue de Huancavelica." In *Colección de memorias científicas, agrícolas é industriales publicadas en distintas épocas.* Vol. II. Mariano Eduardo de Rivero y Ustáriz, Ed. Brussels: Imprenta de H. Goemaere, 1857, 85–176.

NA. "Memorial y relación de las minas de azogue del Pirú." In *Coleción de documentos ineditos, relativos al descubrimiento, conquista y organización de las antiguas posesiones españolas de América y Oceanía, sacados de los Archivos del Reino, y muy especialmente del de Indias.* Vol. VIII. Madrid: Imprenta de Frias y Compañía, 1867, 422–449.

NA. "Pareceres de los Padres de la Compañía de Jesús de Potosí. 1610." In *Pareceres juridicos en asuntos de indias (1601–1718).* Ruben Vargas Ugarte, Ed. Lima: CIP, 1951, 116–131.

NA. "Tensiones y pleitos en una doctrina de naturales." Lima, October 30, 1772. In Comite Arquidiocesano del bicentenario de Túpac Amaru. *Túpac Amaru y la Iglesia: Antología.* Lima: Edubanco, 1983.

Natteri, Oscar. "EMAPA insiste en donacion de terreno para planta." In *Correo del Sur* January 21, 2015. Accessed January 21, 2015, http://diariocorreo.pe/ciudad/emapa-insiste-en-donacion-de-terreno-para-planta-558976/.

Navarra y Rocaful, Duque de la Palata, Melchor. "Relacion del estado del Perú." In *Memorias de los virreyes que han gobernado el Perú durante el tiempo del coloniaje español.* Vol. II. M .A. Fuentes, Ed. Lima: Librería Central de Felipe Bailly, 1859.

Noe, Frank. "Peru." In *Minerals Yearbook Area Reports: International 1969.* Vol. IV. Washington, DC: Bureau of Mines, 1969, 571–585.

Oñate, Pedro de. "Parecer del P. Pedro de Oñate sobre las Minas de Huancavelica 1629." In *Pareceres juridicos en asuntos de indias (1601–1718).* Ruben Vargas Ugarte, Ed. Lima: CIP, 1951, 140–153.

Parés y Franqués, José. *Catástrophe morboso de las minas mercuriales de la villa de Almadén del azogue (1778).* Alfredo Menéndez Navarro, Ed. Cuenca, Spain: Ediciones de la Universidad de Castilla – La Mancha, 1998.

Pennington, J.W. and Gertrude N. Greenspoon. "Mercury." In *Minerals Yearbook: Metals and Minerals (except fuels), 1956.* Washington, D.C.: Bureau of Mines, 1958, 813–830.

Pennington, J.W. and Gertrude N. Greenspoon. "Mercury." In *Minerals Yearbook: Metals and Minerals (except fuels)*, 1958. Vol. I. Washington, D.C.: Bureau of Mines, 1959, 749–765.

Raimondi, Antonio. *El Perú*. 6 Vols. Lima: Imprenta del Estado, 1874.

Raimondi, Antonio. "Huancavelica y mina de azogue, año 1862." In *Notas de viajes para su obra "El Peru."* Vol. III. Lima: Imprenta Torres Aguirre, 1945: 276–289.

Raimondi, Antonio, and M.F. Paz Soldán. *Plano Topografico De La Ciudad De Huancavelica*. Lima: Libreria de Augusto Durand, 1865.

Ramírez, Balthasar, "Descripción del reyno del Pirú del sitio, temple, provincias, obispados y ciudades; de los naturales, de sus lenguas y traje." In *Juicio de límites entre el Perú y Bolivia*. Vol. I. Víctor Maurtua, Ed. Barcelona: Imprenta de Heinrich y Compañía, 1906, 281–363.

Ribera, Pedro de and Antonio de Chaves y de Guevara. "Relación de la Ciudad de Guamanga y sus terminos. Año de 1586." In *Relaciones geográficas de Indias*. Vol. I. Marcos Jiménez de Espada, Ed. Madrid: Ediciones Atlas, 1965, 181–204.

Riva Aguero, José de la, "Descripción anónima del Perú y de Lima a principios del siglo XVII." In *Revista Historica*, Vol. II1, (Lima, 1954): 9–36.

Rivero y Ustariz, Mariano E. de. *Memoria Sobre El Rico Mineral De Azogue De Huancavelica*. Lima: J.M. Masías, 1848.

Salinas y Córdoba, Fray Buenaventura. *Memorial de las Historias del nuevo mundo Pirú*. Lima: Universidad Nacional Mayor de San Marcos, 1957.

San Miguel, Garci Diez de. *Visita hecha a la provincia de Chucuito por Garci Diez de San Miguel en el año de 1567*. Waldemar Espinosa Soriano, Ed. Lima: Casa de la Cultura del Perú, 1964.

Sarmiento de Sotomayor, Conde de Salvatierra, Diego García. "Relación del estado en que deja el gobierno de estos reinos el Conde de Salvatierra al Sr. Virrey Conde de Alba de Liste." In *Colección de las memorias o relaciones que escribieron los virreys del Perú*. Vol. II. Ricardo Beltrán y Rózpide, Ed. Madrid: Imprenta del Asilo de Huérfanos del S.C. de Jesús, 1921, 210–301.

Sebastián, Juan et al. "Parecer de los PP. de la Compañía de Jesús, Juan Sebastián, Esteban de Avila, Manuel Vásquez, Juan Pérez Menacho y Francisco de Vitoria, dado al Virrey D Luis de Velaso, sobre si es lícito repartir indios a las minas que de nuevo se descubrieren. 1599." In *Pareceres juridicos en asuntos de indias (1601–1718)*. Ruben Vargas Ugarte, Ed. Lima: CIP, 1951, 89–93.

Solórzano Pereira, Juan de. *Política Indiana*. Vols. I, IV. Madrid: Ediciones Atlas, 1972.

Strauss, Lester. "Quicksilver at Huancavelica, Peru." In *Mining and Science Press*. Vol. 99 (October 23, 1909): 561–566.

Toledo, Francisco de. "Memorial que D. Francisco de Toledo dió al Rey nuestro señor, del estado en que dejó las cosas del Perú, después de haber sido en él Virrey y Capitán General trece años, que comenzaron en 1569." In *Colección de las memorias o relaciones que escribieron los virreys del Perú*. Vol. I. Ricardo Beltrán y Rózpide, Ed. Madrid: Imprenta del Asilo de Huérfanos del S.C. de Jesús, 1921, 71–107.

Toledo, Francisco de. "Ordenanzas que el Señor Viso Rey Don Francisco Toledo hizo para el buen gobierno de estos Reynos del Perú y Repúblicas de él." In *Relaciones de los Virreyes y Audiencias que han Gobernado el Perú*. Vol. I. Sebastian Lorente, Ed. Lima: Imprenta del Estado, 1867, 33–366.

Toledo y Leiva, Manuel. "Parecer del P. Manuel Toledo y Leiva, Rector del Colegio de la Compañía de Jesús de Huancavelica sobre la Mita de Potosí, a petición del Sr. D.D. José Santiago Concha Oidor de Lima y Gobernador de Huancavelica, en virtud de R.C. expedida en Madrid el 6 de Diciembre de 1719." In *Pareceres jurídicos en asuntos de indias (1601–1718)*. Ruben Vargas Ugarte, Ed. Lima: CIP, 1951, 168–183.

Torres de Portugal, Conde de Villardompardo, Fernando de. "Memoria gubernativa del Conde del Villardompardo." Lima, May 25, 1592 or 1593. In *Los virreyes españoles en America durante el gobierno de la casa de Austria. Perú*. Vol. I. Lewis Hanke and Celso Rodríguez, Eds. Madrid: IMNASA, 1978, 203–250.

Tschudi, Juan Jacobo von. *El Perú. Esbozos de viaje realizados entre 1838 y 1842*. Peter Kaulicke, Ed. Lima, Pontífica Universidad Católica del Perú, 2003.

Ulloa, Antonio de. *Noticias americanas: Entretenimientos phisicos-históricos, sobre la América Meridional y la Septentrianal Oriental*. Granada, Spain: Universidad de Granada, 1992.

Ulloa, Antonio de. *Viaje a la América meridional*. 2 Vols. Andrés Samuell, Ed. Madrid: Historia 16, 1990.

Velasco, Luis de. "Relación del Sr. Virrey, D. Luis de Velasco, al Sr. Conde de Monterrey sobre el estado del Perú." In *Colección de las memorias o relaciones que escribieron los virreys del Perú*. Vol. I. Ricardo Beltrán y Rózpide, Ed. Madrid: Imprenta del Asilo de Huérfanos del S.C. de Jesús, 1921, 108–140.

Villaba, Victorián de. "Vista del fiscal Victorián de Villaba, sobre los abusos de la mita." In *Vida y obra de Victorián de Villaba*. Ricardo Levene, Ed. Buenos Aires: Instituto de Investigaciones Históricas, 1946, lvi–lxiv.

Wessel, F.W. "Peru." In *Minerals Yearbook Area Reports: International 1972*. Vol. III. Washington, DC: Bureau of Mines, 1972, 639–648.

Wessel, F.W. "Peru." In *Minerals Yearbook Area Reports: International 1974*. Vol. III. Washington, DC: Bureau of Mines, 1974, 717–728.

Witt, Heinrich. *Diario y observaciones sobre el Perú (1824–1890)*. Kika Garland de Montero, Trans. Lima: Corporación Financiera de Desarrollo, S.A., 1987.

Zubilete, Raúl. "En chakus logran más de una tonelada de fibra de vicuña." In *Correo del Sur*. January 1, 2015. Accessed on January 1, 2015. http://diariocorreo.pe/ciudad/logran-mas-de-una-tonelada-de-fibra-554480/.

Secondary Sources

Abrines, María C.N. "El gobierno de Carlos de Beranger en Huancavelica (1764–1767)." In *Jahrbuch Für Geschichte Von Staat, Wirtschaft Und Gesellschaft Lateinamerikas.* Vol. III4 (1997): 105–126.

Acosta, Luis Vilma Milletich and Enrique Tandeter. "El comercio de efectos de la tierra en Potosí. 1780–1810." In *Minería colonial Latinoamericana.* Dolores Avila, Inés Herrera and Rina Ortíz, Eds. Mexico City: Instituto Nacional de Antropología e Historia, 1992, 137–153.

Alcser, Kirsten, et al. "Occupational Mercury Exposure and Male Reproductive Health." In *American Journal of Industrial Medicine.* Vol. I5, No. 5 (1989): 517–529.

Allison, Marvin. "Paleopathology in Peru." In *Natural History,* Vol. VIII8, No. 2 (1979): 74–82.

Andersen, A, et al. "A neurological and neurophysiological study of chloralkali workers previousy exposed to mercury vapor." In *Acta Neurologica Scandinavica,* Vol. VIII8, No. 6, (December 1993): 427–433.

Arana, Pedro, *Las minas de azogue del Perú.* Lima: Imprenta de "El Lucero," 1901.

Arduz, Eguía Gastón. *Ensayos sobre la historia de la minería altoperuana.* Madrid: Paraninfo, 1985.

Assadourian, C. Sempat. *El sistema de la economia colonial.* Mexico City: Editorial Nueva Imagen, 1983.

Assadourian, C. Sempat. "La crisis demográfica del siglo XVI y la transición del Tawantinsuyo al sistema mercantil colonial." In *Población y mano de obra en América Latina,* Nicolás Sánchez Albornoz, Ed. Madrid: Alianza Editorial, 1985, 69–93.

Assadourian, C. Sempat, Heraclio Bonilla, Antonio Mitre and Tristan Platt. *Minería y espacio económico en los Andes. Siglos XVI-XX.* Lima: Instituto de Estudios Peruanos, 1980.

Aste, Daffos J. *Minería Y Desarrollo Regional: Los Casos De Junín Y Huancavelica, 1970–86.* Lima, Perú: Fundación Friedrich Ebert, 1989.

Attman, Arthur. *American Bullion in the European World Trade, 1600–1800.* Eva and Allan Green, Trans. Goteborg, Sweden: Kungl. Vetenskaps – och Vitterhets-Samhallet i Goteborg, 1986.

Ayala, Javier de. "Estudio preliminar." In *Servidumbres personales de indios.* Javier de Ayla, Ed. Sevilla: Escuela de Estudios Hispanos-Americanos, 1946, xi–xxvii.

Bakewell, Peter. *Miners of the Red Mountain: Indian Labor in Potosí, 1545–1650.* Albuquerque: University of New Mexico Press, 1984.

Bakewell, Peter. "Registered Silver Production in the Potosi District, 1550–1735." In *Jahrbuch fur Geschichte von Staat, Wirtschaft und Gesellschaft Lateinamerikas,* Vol 12. (1975): 67–103.

Bakewell, Peter. "Technological Change in Potosi: The Silver Boom of the 1570s." In *Mines of Silver and Gold in the Americas*. Peter Bakewell, Ed. Brookfield, VT: Variorum, 1997.

Bakir, F., et al. "Methylmercury Poisoning in Iraq." In *Science*, Vol. I81, No. 4096 (July 20, 1973): 230–241.

Balaan, Marvin and Daniel Banks. "Silicosis." In *Environmental and Occupational Medicine*. 3rd Ed. William Rom, Ed. New York: Lippincott-Raven Publishers, 1998, 435–448.

Bargallo, Modesto. La amalgamación de los minerales de plata en hispanoamerica colonial. Mexico City: Compañia Fundidor de Fierro y Acero de Monterrey, 1969.

Bargallo, Modesto. *La minería y metalurgía en la América española durante la época colonial*. Mexico City: Fondo de Cultura Económica, 1955.

Barlow, S.M., et al. "Reproductive hazards at work." In *Hunter's Diseases of Occupations*. 8th ed. P.A.B. Raffle et al., Eds. London: E. Arnold, 1994, 723–742.

Barnadas, Josep *Charcas: origines históricos de una sociedad colonial*. La Paz: Centro de Investigación y Promoción del Campesinado, 1973.

Basadre, Jorge. *El Conde de Lemos y su tiempo*. Lima: Editorial Huascaran, S.A., 1948.

Basadre, Jorge. "El Régimen de la Mita." In *Letras*, Vol. III (Lima: 1937): 325–364.

Basto Girón, Luis J. "Las mitas de Huamanga y Huancavelica." In *Perú Indígena*, Vol. V, No. 13 (December, 1954): 215–242.

Basto Girón, Luis J. *Salud y enfermedad en el campesino peruano del siglo XVII*. Lima: Instituto de Etnología y Arqueología, Universidad Nacional Mayor de San Marcos, 1957.

Berry, Edward and Joseph T. Singewald. *The Geology and Paleontology of the Huancavelica Mercury District*. Baltimore: Johns Hopkins Press, 1922.

Bidstrup, P. Lesley. *Toxicology and Mercury and its Compounds*. New York: Elsevier Publishing Company, 1964.

Bonilla, Heraclio. "1492 y la población indígena de los Andes." In *Los conquistados: 1492 y la poblacion indígena de las Américas*. Heraclio Bonilla, Robin Blackburn, et al., Eds. Bogotá: Tercer Mundo Editores/ Facultad latinoamericana de ciencias sociales, 1992, 103–125.

Bonilla, Heraclio. "Religious Practices in the Andes and their Relevance to Political Struggle and Development." In *Mountain Research and Development* Vol. II6, No.4 (2006): 336–342.

Bonilla, Heraclio and Carmen Salazar. *Formación del mercado laboral para el sector minero. (La experiencia de Huancavcelica, Perú, 1950–1978)*. Lima: PUCP, 1983.

Bradby, Barbara. "The 'Black Legend' of Huancavelica: The mita debates and opposition to wage-labor in the colonial mercury mine." In *Hombres, técnica, plata: minería y sociedad en Europa y América, siglos XVI-XIX.* Julio Sánchez Gómez and Guillermo Mira Delli-Zotti, Eds. Seville: Aconcagua Libros, 2000, 227–257.

Brading, D.A. and Harry E. Cross. "Colonial Silver Mining: Mexico and Peru." In *The Hispanic American Historical Review*, Vol. V2, No. 4 (November, 1972): 545–579.

Brooks, William E, Schworbel, Gabriela, and Castillo, Luis Enrique. "Amalgamation and Small-scale Gold Mining in the Ancient Andes." In *Mining and Quarrying in the Ancient Andes: Sociopolitical, Economic and Symbolic Dimensions.* Nicholas Tripcevich and Kevin Vaughan, Editors. New York: Springer Publishing, 2013, 213–229.

Brooks, William E., Estevan Sandoval, Miguel Yépez and Howell Howard, *Peru Mercury Inventory, 2006.* Washington, DC: United States Geological Service, 2007.

Brown, Kendall. "El ingeniero Pedro Subiela y el desarrollo tecnológico en las minas de Huancavelica (1786–1821)." In *Histórica.* Vol. IIIo, No.1 (July, 2006): 165–184.

Brown, Kendall. "La crisis financiera peruana al comienzo del siglo XVIII, la minería de plata y la mina de azogue de Huancavelica." In *Revista de Indias*, Vol. XLVIII, Nos. 182–183 (1988): 349–381.

Brown, Kendall. "La distribución del mercurio a finales del periodo colonial, y los trastornos provocados por la independencia hispanoamericana." In *Minería colonial Latinoamericana*, Dolores Avila, Inés Herrera and Rina Ortíz, Eds. Mexico City: Instituto Nacional de Antropología e Historia, 1992, 155–160.

Brown, Kendall. "La recepción de la tecnología minera española en las minas de Huancavelica, siglo XVIII." In *Saberes andinos: ciencia y tecnología en Bolivia, Ecuador y Perú.* Marcos Cueto, Ed. Lima: Instituto de Estudios Peruanos, 1995.

Brown, Kendall. "Los cambios tecnológicos en las minas de Huancavelica, siglo XVIII." In *Hombres, técnica, plata: minería y sociedad en Europa y América, siglos XVI-XIX.* Julio Sánchez Gómez and Guillermo Mira Delli-Zotti, Ed. Seville: Aconcagua Libros, 2000.

Brown, Kendall. "Nordenflicht, Thaddeus von. Tratado del arreglo y reforma que conviene introducir en la minería del reino del Perú para su prosperidad, conforme al sistema y práctica de las naciones de Europa más versadas en este ramo, presentado de oficio al superior gobierno de estos reinos por el barón de Nordenflicht. Estudio preliminar de José Ignacio López Soria." In *Histórica* 31.1 (July, 2007): 213.

Brown, Kendall. "Workers' Health and Colonial Mercury Mining at Huancavelica, Peru." In *The Americas*, Vol. V7. No. 4 (April, 2001): 467–496.

Buechler, Rose M. *The Mining Society of Potosi. 1776–1810.* Syracuse: Syracuse University Department of Geography, 1981.

Buechler, Rose M. "Technical Aid to Upper Peru: The Nordenflicht Expedition." In *Journal of Latin American Studies*, Vol. V, No. 1 (1973): 37–77.

Burger, R and R. Matos. "Atalla: A Center on the Periphery of the Chauvín Horizon." In *Latin American Antiquity* Vol. I3, No. 2 (2002): 10–25.

Bury, Jeffrey. "Livelihoods in Transition: Transnational Gold Mining Operations and Local Change in Cajamarca, Peru." *Geographic Journal*, Vol. I70, No. 1 (March, 2004), 78–91.

Cabrera La Rosa, Augusto, "Situación actual de la minería del mercurio en el Perú." In *Minería y Metalurgia: Boletín official de minas, metalurgia y combistibles* (Madrid: Veritas, 1954), 3–12.

Cahill, David. "Curas and Social Conflict in the Doctrinas of Cuzco, 1780 1814." In *Journal of Latin American Studies*. Vol. I6, No. 2 (November, 1984): 241–276.

Cañedo-Arguelles Fábrega, Teresa. "Efectos de Potosí sobre la población indígena del Alto Perú. Pacajes a mediados del siglo XVII." In *Revista de Indias*. Vol. XLVIII, Nos. 182–183 (1988): 237–255.

Cañedo-Arguelles Fábrega, Teresa. *Potosí: La versión aymará de un mito europeo. La minería y sus efectos en las sociedades andinas del siglo XVII (La Provincia de Pacajes)*. Madrid: Editorial Catriel, 1993.

Carrasco, Tulio. *Cronología de Huancavelica (Hechos, poblaciones y personas)*. Lima: Companía de Minas Buenaventura, 2003.

Castañeda Delgado, Paulino."El tema de las minas en la ética colonial española." In *La mineria hispana e iberoamericana*. Vol. I. NA. León, Spain: Catedra de San Isidoro, 1970, 333–354.

Castañeda Delgado, Paulino. "Un capítulo de ética Indiana española: los trabajos forzados en las minas." In *Anuario de Estudios Americanos*, Vol. XXVII, (1970): 815–916.

Celestino, Olinda, and Albert Meyers. *Las Cofradías en el Perú: region central*. Frankfurt: Vervuert, 1981.

Cherry, N.M. "Neurotoxic effects of workplace exposures." In *Hunter's Diseases of Occupations*. P.A.B. Raffle et al., Eds. 8th ed. London: E. Arnold, 1994, 75–89.

Choque Canqui, Roberto. "El papel de los capitanes de indios de la provincia de Pacajes 'en el entero de la mita' de Potosí." In *Revista Andina*, Vol. I, No. 1 (September, 1983): 117–125.

Cipolla, Carlo M. *Conquistadores, piratas, mercaderes: La saga de la plata española*. Ricardo González, Trans. Buenos Aires: Fondo de Cultura Económica de Argentina, 1999.

Clark, G.N. "The Early Modern Period." In *The European Inheritance*. Earnest Barker et al, Eds. Vol. II. London: Clarendon Press, 1954, 3–181.

Cobb, Gwendolyn. *Potosí y Huancavelica, bases económicas, 1545–1640*. La Paz: Banco Minero de Bolivia, 1977.

Cobb, Gwendolyn. "Supply and Transportation for the Potosi Mines, 1545–1640." In *Hispanic American Historical Review*, Vol. II9, No. 1. (1949): 25–45.

Cole, Jeffrey A. *The Potosí Mita, 1573–1700. Compulsory Indian Labor in the Andes*. Stanford: Stanford University Press, 1985.

Collinson, Helen, Ed., *Green Guerrillas: Environmental Conflicts and Initiatives in Latin America and the Caribbean*. London: Latin American Bureau, 1996.

Contreras, Carlos. "El reemplazo del beneficio de patio en la minería peruana, 1850–1913." In *Revista de Indias*, Vol. V9, No. 216 (1999): 391–416.

Contreras, Carlos. *La ciudad del mercurio: Huancavelica, 1570–1700*. Lima: Instituto de Estudios Peruanos, 1982.

Contreras, Carlos, and Ali Díaz. *Los intentos de reflotamiento de la mina de azogue de Huancavelica en el siglo XIX*. Lima, NP, 2007.

Cook, Noble David. *Born to Die: Disease and New World Conquest, 1492–1650*. Cambridge: Cambridge University Press, 1998.

Cook, Noble David. *Demographic Collapse: Indian Peru, 1520–1620*. Cambridge: Cambridge University Press, 1981.

Cooke Colin, Prentiss Balcom, Harald Biester, and Alexander Wolfe. "Over three millennia of mercury pollution in the Peruvian Andes." In *Proceedings of the National Academy of Sciences*. Vol. I02, No. 22 (June, 2009): 8830–8834.

Cooke, Colin, Holger Hintelmann, Jay Ague, Richard Burger, Harald Biester, Julian Sachs and Daniel Engstgrom. "Use and legacy of mercury in the Andes." In *Environmental Science and Technology*. Vol. IV7, No. 9 (May, 2013): 4181–4188.

Cortés Salinas, María del Carmen. "Una polémica en torno a la mita de Potosí a fines del siglo XVIII." In *Revista de Indias*. Vol. XXX, Nos. 119–122 (January–December, 1970):131–215.

Crespo Rojas, Alberto. "El reclutamiento y los viajes en la 'mita' del cerro Potosí." In *La mineria hispana e iberoamericana*. Vol. I. N.A., León, Spain: Catedra de San Isidoro, 1970, 467–482.

Crespo Rojas, Alberto. *La guerra entre vicuñas y vascongados (Potosí, 1622–1625)*. Lima: Tipografía Peruana, 1956.

Crespo Rojas, Alberto. "La 'mita' de Potosí." In *Revista Histórica*, Vol. II2 (Lima, 1955–56): 169–182.

Crosby, Alfred. *The Columbian Exchange: Biological and Cultural Consequences of 1492*. Westport CT: Greenwood Press, 1972.

Cross, Harry E. "South American Bullion Production and Export 1550–1750." In *Precious Metals in the Later Medieval and Early Modern Worlds*. J.F. Richards, Ed. Durham, NC: Carolina Academic Press, 1983.

Delgado de Castro, Raquel. *El Despertar De Huancavelica*. Lima: C.A. Castrillón, 1927.

Dell, Melissa. "Persistent Effgects of Peru's Mining Mita." In *Essays in Economic Development and Political Economy*. Diss. Massachusetts Institute of Technology, 2012.

Deustua, José. *The Bewitchment of Silver: The Social Economy of Mining inNineteenth-Century Peru*. Athens: Ohio University Press, 2000.

D'Itri, Patricia and Frank D'Itri. *Mercury Contamination: A Human Tragedy*. New York: John Wiley and Sons, 1977.

Dobado González, Rafael. "Las minas de Almadén, el monopolio del azogue y la producción de plata en Nueva España en el siglo XVIII." In *La savia del imperio: tres estudios de economía colonial*. Julio Sánchez Gómez, Guillermo Mira Delli-Zotti and Rafael Dobado, Eds. Salamanca, Ediciones Universidad Salamanca, 1997, 401–471.

Dobado González, Rafael. "Salarios y condiciones de trabajo en las minas de Almaden, 1758–1839." *In La economía española al final del Antiguo Régimen. II. Manufacturas*. Pedro Tedde, Ed. Madrid: Alianza Editorial/ Banco de España, 1982, 337–438.

Dobyns, Henry F. "An Outline of Andean Epidemic History to 1720." In *Bulletin of the History of Medicine*, Vol. III7, No. 6 (November–December, 1963): 493–515.

Doña Nieves, Francisco. "Trabajo y salud en las minas de plata americanas del siglo XVI." In *Anales de la Real Academia de Medicina y Cirugia de Cádiz*, Vol. II8, No. 1 (1992): 271–281.

Duviols, Pierre. *La destrucción de las religions andinas (Conquista y colonia)*. Mexico: UNAM, 1977.

Ecos Lima, Máximo Enrique. "Informe de monitoreo agua de consumo humano 'Distritos de Huancavelica, Izcuchaca y Palca'" Lima: Companía de Minas Buenaventura, 2010.

Eguren, Mariana, Carolina de Belaunde and Ana Luisa Burga. *Huancavelica cuenta: temas de historia huancavelicana contados por sus protagonistas*. Lima: Instituto de Estudios Peruanos, 2005.

Ehrenberg, Richard, et al. "Effects of Elemental Mercury Exposure at a Thermometer Plant." In *American Journal of Industrial Medicine* Vol. I9 (1991): 495–507.

Enock, C. Reginald. *The Andes and the Amazon: Life and Travel in Peru*. London: T. Fisher Unwin, 1907.

Espinoza Flores, Mariela. *Huancavelica: Rincón De Misterios Y Encantos/ A Spot Full of Mysteries and Charm*. Lima: Compañía de Minas Buenaventura, 2009.

Espinoza Gonzales, Rubén Darío. "Una vision de la Arqueología de Huancavelica." In Arqueología y Desarrollo: Experiencias y posibilidades en el Perú. Luis Valle Alvarez, Editor. Trujillo, Peru: Ediciones SIAN, 2010, 67–78.

Evans, Brian M. "Census Enumeration in Late Seventeenth-Century Alto Perú: The Numeración General of 1683–1684." In *Studies in Spanish Population History*. D.J. Robinson, Ed. Boulder: Westview Publishers, 1981, 25–44.

Evans, Hugh. "Mercury." In *Environmental and Occupational Medicine*. 3rd. ed. William Rom, Ed. New York: Lippincott- Raven Publishers, 1998. 997–1003.

Evans, Hugh, et al. "Behavioral effects of mercury and methylmercury." In *Federation Proceedings*, Vol. III4, No. 9 (August, 1975): 1858–1867.

Ezquerra Abadia, Ramón, "Problemas de la mita de Potosí en el siglo XVIII." In *Lamineria hispana e iberoamericana*. Vol. I. N.A. Leon, Spain: Catedra de San Isidoro, 1970, 483–511.

Falnoga, I.M. Tusek-Znidaric, et al. "Mercury, Selenium, and Cadmium in Human Autopsy Samples from Idrija Residents and Mercury Mine Workers." In *Environmental
Research*, Volume 84, Section A, (2000): 211–218.

Favre, Henri. "Evolución y situación de las haciendas en la región de Huancavelica,
Perú." In *Revista del Museo Nacional*. Vol. III3 (1964): 237–257.

Favre, Henri. "La industria minera de Huancavelica en la década de 1960." In *Boletín de
Lima*. No. 161 (2010): 85–89.

Fernández Alonso, Serena. "Los mecenas de la plata: el respaldo de los virreyes a la
actividad minera colonial en las primeras decadas del siglo XVIII. El gobierno del
Marques de Casa Concha en Huancavelica, 1723–1726." In *Revista De Indias*. Vol. VIo,
No. 219 (Madrid: 2000): 345–371.

Fisher, John A. *Silver Mines and Silver Miners in Colonial Peru, 1776–1824*. Monograph
series No. 7. Liverpool: Center for Latin American Studies, University of Liverpool,
1977.

Fisher, John A.. "Silver Production in the Viceroyalty of Peru, 1776–1824." In *Mines of
Silver and Gold in the Americas*. Peter Bakewell, Ed. Brookfield, VT: Variorum, 1997.

Flynn, Dennis. "A New Perspective on the Spanish Price Revolution: The Monetary Approach to the Balance of Payments." In *World Silver and Monetary History in the 16th
and 17th Centuries*. Brookfield, VT: Variorum, 1996, 388–406.

Flynn, Dennis. "Fiscal Crisis and the Decline of Spain (Castile)". In *World Silver and
Monetary History in the 16th and 17th Centuries*. Brookfield, VT: Variorum, 1996,
139–147.

Flynn, Dennis and Arturo Giraldez. "China and the Manila Galleons." In *World Silver
and Monetary History in the 16th and 17th Centuries*. Brookfield, VT: Variorum, Brookfield, VT: Variorum, 1996, 71–90.

Fox, K.V. "Pedro Muñiz, Dean of Lima Cathedral and the Indian Labor Question (1603)."
In *Hispanic American Historical Review*. Vol. IV2, No. 1 (1962): 63–88.

Fox, Richard and Orin Starn, Eds. *Between Resistance and Revolution: Cultural Politics
and Social Protest*. New Brunswick: Rutgers University Press, 1997.

Frith, John. "Syphilis: Its Early History and Treatment Until Penicillin, and the Debate
on its Origins." In *Journal of Military and veterans Health*. Vol. IIo, No. 4 (November,
2012): 49–58.

Fuentes Bajo, María Dolores. "El azogue en las postrimerías del Perú colonial." In *Revista de Indias*. Vol. XLVI (January-June, 1986): 75–105.

Galeano, Eduardo. *Las venas abiertas de América Latina*. Mexico City: Ediciones Siglo
Veintiuno, 1971.

Gibson, Charles. "Introduction." In *The Black Legend: Anti-Spanish Attitudes in the Old
World and the New*. Charles Gibson, Ed. New York: Alfred A. Knopf, 1971, 3–27.

Goldwater, Leonard J. *Mercury: A History of Quicksilver*. Baltimore: York Press, 1972.

Golte, Jurgen. *Repartos y rebeliones: Túpac Amaru y las contradicciones de la economia colonial.* Lima: Instituto de estudios Peruanos, 1980.

Gómez Rivas, León, *El virrey del Perú Don Francisco de Toledo.* Madrid: Instituto Provincial de Investigaciones y Estudios Toledanos, 1994.

Gonzales, Gustavo F., Vilma Tapia, Manuel Gasco and Carlos Carillo. "Hemoglobina maternal en el Perú: Diferencias regionales y su asociación con resultados adversos perinatales." In *Revista Peruana de Medicina Experimental y Salud Pública.* Vol. II8, No. 3 (2011): 484–491.

Griffiths, Nicholas. *The Cross and the Serpent: Religious Repression and Resurgence in Colonial Peru.* Norman: University of Oklahoma Press, 1996.

Gutiérrez Brockington, Lolita. *Blacks, Indians and Spaniards in the Eastern Andes: Reclaiming the Forgotten in Colonial Mizque, 150–1782.* Nebraska University Press, 2006.

Haan, Stef de. *Catálogo De Variedades De Papa Nativa De Huancavelica, Perú.* Lima: Centro Internacional de la Papa, 2006.

Haan, Stef de, and Henry Juárez. "Land Use and Potato Genetic Resources in Huancavelica, Central Peru." In *Journal of Land Use Science.* Vol. V, No. 3 (2010): 179–195.

Haan, Stef de, Jorge Nuñez, Merideth Bonierbale and Marc Ghislain. "Multilevel Agrobiodiversity and Conservation of Andean Potatoes in Central Peru: Species, Morphological, Genetic, and Spatial Diversity." In *Mountain Research and Development.* Vol. IIIo, No. 3 (2010): 222–231.

Hadley, William. "Report on the quicksilver mines of Huancavelica." In *Boletín del Ministerio de Fomento.* Vol. II, No. 1 (Lima, 1904):43–44.

Hagan, Nicole, Nicholas Robins, Heileen Hsu-Kim, Susan Halabi, Rubén Darío Espinoza Gonzales, Daniel Richter and John Vandenberg. "Residential Mercury Contamination in Adobe Brick Homes in Huancavelica, Peru." In *PLoS ONE* Vol. VIII No. 9. (September, 2013): 1–9.

Hamilton, Earl. *American Treasure and the Price Revolution in Spain, 1501–1650.* Cambridge: Harvard University Press, 1934.

Hanke, Lewis. "The Social History of Potosi." In *La minería hispana e iberoamericana,* Vol. I. N.A. León, Spain: Cátedra de San Isidoro, 1970: 451–465.

Hanninen, Helena. "Behavioral Effests of Occupational Exposure to Mercury and Lead." In *Acta Neurologica Scandinavica,* Vol. VI6, Supplement 92 (1982): 167–173.

Harrison, Regina. "The Theology of Concupiscence: Spanish-Quechua Confessional Manuals in the Andes." In *Coded Encounters: Writing, Gender, and Ethnicity in Colonial Latin America.* Francisco Javier Cevallos-Candau, Ed. Amherst: University of Massachusetts Press, 1994.

Hawley, C.E. Notes on the Quicksilver Mine of Santa Barbara, Peru. In *American Journal of Science.* 2nd series. Vol 45 (1868): 5–9.

Hernández, José A. "Antiguedad y actualidad de la Villa Rica de Oropesa." In *Peruanidad*. Vol. III, No. 12 (Lima, 1943): 942–944.

Hindery, Derrick. *From Enron to Evo: Pipeline Politics, Global Environmentalism, and Indigenous Rights in Bolivia*. Tucson: University of Arizona Press, 2013.

Hunefeldt, Christine. "Comunidad, curas y comuneros hacia fines del periodo colonial: ovejas y pastores indomados en el Perú." In *Hisla*, No. 2, (1983): 3–31.

Hursh, John, et al. "Clearance of Mercury (Hg-197, Hg-203) Vapor Inhaled by Human Subjects." In *Archives of Environmental Health* Vol. III1 (Nov/Dec 1976): 302–309.

Hursh, John, et al., "The Effect of Ethanol on the Fate of Mercury Vapor Inhaled by Man." In *The Journal of Pharmacology and Experimental Therapeutics*, Vol. II14, No.3 (1980): 520–527.

Iwata, Toyoto, Mineshi Sakamoto, et al. "Effects of mercury vapor exposure on neuromotor function in Chinese miners and smelters." In *International Archive of Occupational Environmental Health*. Vol. VIII0, (2007): 381–387.

Jakob, W. "Sumario de las ordenanzas mineras del Perú." In *Revista del Instituto de Historia del derecho Ricardo Levene*, Vol. III (Buenos Aires, 1972): 273–288.

Jara, Alvaro. *Tres ensayos sobre economía minera hispanoamericana*. Santiago: Universidad de Chile, 1966.

Jennings, Justin. "Inca Imperialism, Ritual Change, and Cosmological Continuity in the Cotahuasi Valley of Peru." In *Journal of Anthropological Research* Vol. V9, No. 4 (2003): 433–462.

Kark, R.A. "Clinical and neurochemical aspects of inorganic mercury intoxication." In *Handbook of Clinical Neurology*. P. Vinken and G. Bruyn, Eds. Vol. III6. New York: Elsevier, 1979, 147–197.

Kishi, Reiko, et al. "Subjective Symptoms and Neurobehavioral Performances of Ex-Mercury Miners at an Average of 18 years After the Cessation of Chronic Exposure to Mercury Vapor." In *Environmental Research*. Vol. VI2, No. 2, (1993): 289–302.

Kishi, Reiko, Rikuo Doi, et al. "Residual neurobehavioural effects associated with chronic exposure to mercury vapour." In *Occupational and Environmental Medicine*. Vol. V1, No. 1 (January, 1994): 35–41.

Kudsk, Nielson. "The Influence of Ethyl Alcohol on the Absorption of Mercury Vapour from the Lungs in Man." In *Acta Pharmacologia et* Toxicologia. Vol. II3 (1965): 263–274.

Lacerda, L.D. "Global mercury emissions from gold and silver mining." In *Water, Air and Soil Pollution*. Vol. IX7 (1997): 209–221.

Lacerda, L.D. and R.V. Marins. "Anthropogenic mercury emissions to the atmosphere in Brazil: The impact of gold mining." In *Journal of Geochemical Exploration*. Vol. V8 (1997): 223–229.

Lang, Mervyn. "El derrumbe de Huancavelica en 1786. Fracaso de una reforma bourbónica." In *Histórica*. Vol. I0, No. 2 (Dec., 1986): 213–226.

Larco Hoyle, Rafael. *Los Mochicas*. Vol. II. Lima: Museo Arqueológico Rafael Larco Herrera, 2001.

Lewin, Boleslao, *La rebelión de Túpac Amaru y los orígines de la independencia de Hispanoamérica*. Buenos Aires: Libreria Hachette, 1957.

Lichtenstein, G., Baldi, R., Villalba, L., Hoces, D., Baigún, R. & Laker, J. 2008. "Vicugna vicugna." In *The IUCN Red List of Threatened Species*. Version 2014.3. www.iucnredlist.org.

Lockhart, James. *Spanish Peru, 1532–1560. A Social History*. Madison: University of Wisconsin Press, 1994.

Lohmann Villena, Guillermo. "Enrique Garcés, descubridor del mercurio en el Perú, poeta y arbitrista." In *Studia*, Vols. 27–28 (August-December 1969): 7–62.

Lohmann Villena, Guillermo. "La minería en el marco del virreinato peruano: Invenciones, sistemas, técnicas y organización industrial." In *La minería hispana e iberoamericana*, Vol. I. N.A. Leon, Spain: Catedra de San Isidoro, 1970:639–655.

Lohmann Villena, Guillermo. *Las minas de Huancavelica en los siglos XVI y XVII*. Lima: Pontífica Universidad Católica del Perú, 1999.

López Cisneros, Carmen. *Huancavelica Ya No Es Tierra Del Mercurio*. Lima: Fundación Friedrich Ebert, 1983.

Lorente, Sebastián. *Historia del Perú bajo la dinastía Austriaca*. Paris: Imprenta de Poissy, ND.

Magos, Laszlo, et al. "The Depression of Pulmonary Retention of Mercury Vapor by Ethanol: Identification of the Site of Action." In *Toxicology and Applied Pharmacology*, Vol. II6 (1973): 180–183.

Manrique, Nelson. *Colonialismo y pobreza campesina. Caylloma y el valle de Colca, siglos XVI-XX*. Lima: Desco, 1985.

Marichal, Carlos. "The Spanish-American Silver Peso: Export Commodity and Global Money of the Ancien Regime, 1550–1800." In *From Silver to Cocaine: Latin American Commodity Chains and the Building of the World Economy, 1500–2000*. Steven Topik, Carlos Marichal and Zephyr Frank, Eds. Durham, NC: Duke University Press, 2006.

Marsh, David O. "Organic mercury: methylmercury compounds." In *Handbook of Clinical Neurology*. Vol. II6. P. Vinken and G. Bruyn, Eds. New York: Elsevier, 1979, 73–81.

Martin, David Barrett. "The Cause of Death in Smallpox: An Examination of the Pathology Record." In *Military Medicine* Vol. I67, No. 7 (2002): 546–551.

Matilla Tascón, Antonio. *Historia de las minas de Almadén*. Vol II. Madrid: Instituto de Estudios Fiscales/ Minas de Almadén y Arrayanes, 1987.

McKee, E.H., D.C. Noble, and Cesar Vidal. "Timing of Volcanic and Hydrothermal Activity, Huancavelica Mercury District, Peru." In *Economic Geology and the Bulletin of the Society of Economic Geologists*. Vol. VIII1, No. 2 (1986):489–492.

Mendoza Ruíz, Amador. *Crónicas Del Tren Macho*. Lima, Perú: Niger Editions, 1998.

Menéndez Navarro, Alfredo. *Un mundo sin sol. La salud de los trabajadores de las minas de Almadén, 1750–1900*. Granada: Universidad de Granada, 1996.

Millones, Luis. "Religion and Power in the Andes: Idolatrous Curacas of the Central Sierra." In *Ethnohistory*. Vol. II6, No. 3 (Summer, 1979): 243–263.

Mills, Kenneth. *Idolatry and Its Enemies: Colonial Andean Religion and Extirpation, 1640–1750*. Princeton: Princeton University Press, 1997.

Mills, Kenneth."The Limits of Religious Coercion in Mid-Colonial Peru." In *Past and Present*, No. 145 (1994): 84–121.

Mira Delli-Zotti, Guillermo. "El Real Banco de San Carlos de Potosí y la minería Altoperuana colonial, 1779–1825." In *La savia del imperio: tres estudios de economia colonial*. Julio Sánchez Gómez, Guillermo Mira Delli-Zotti and Rafael Dobado, Eds. Salamanca, Ediciones Universidad Salamanca, 1997, 265–399.

Molina Martínez, Miguel. *Antonio de Ulloa en Huancavelica*. Granada, Spain: University of Granada, 1995.

Molina Martínez, Miguel. "Tecnica y laboreo en Huancavelica a mediados del siglo XVIII." In *Europa e Iberoamerica: Cinco siglos de Intercambios*. Vol. II. María Justina Sarabia Viejo, et al., Eds. Seville, Spain: Asociación de Historiadores Latinoamericanistas Europeos/ Consejeria de Cultura y Medio Ambiente, 1992, 395–405.

Mugica, Martín José de. "Abusos. De varias clases de mitas y carácter perezoso del Indio." In "Las mitas de Huamanga y Huancavelica." In *Perú Indígena*. Luis J. Basto Girón. Vol. V, No. 13 (December, 1954): 223–241.

Munares-García, Oscar, Guillermo Gómez-Guizado, and Juan Barboza-Del Carpio. "Niveles de hemoglobina en gestantes atendidas en establecimientos del Ministerio de Salud del Perú, 2011." In *Revista Peruana de Medicina Experimental y Salud Publica*. Vol. II9, No. 3 (2012): 329–336.

Muñoz García, Angel. "Introducción." In *Thesaurus Indicus*. Diego de Avendaño, Angel Muñoz García, Trans. and Eds. Pamplona: Ediciones Universidad de Navarra, 2001, 13–169.

NA. "Health Effects." In *Toxicological Profile for Mercury*. Agency for Toxic Substances and Disease Registry, United States Department of Health and Human Resources. Atlanta, GA: United States Department of Health and Human Resources, 1999, 29–361.

NA. *The Economics of Mercury*. Seventh Edition London: Roskill Information Services, 1990.

Nash, Donna, and Patrick Williams. "Sighting the Apu: A GIS Analysis of Wari Imperialism and the Worship of Mountain Peaks." In *World Archaeology* Vol. II8, No. 3 (2006): 455–468.

Neal, Paul, et al. *Mercurialism and Its Control in the Felt-Hat Industry*. Washington, DC: Government Printing Office, 1941.

Noble, Donald, and Cesar Vidal. "Association of Silver with Mercury, Arsenic, Antimony, and Carbonaceous Material at the Huancavelica District, Peru." In *Economic
Geology*, Vol 85, No. 7 (1990): 1645–1650.

Nriagu, "Mercury pollution from the past mining of gold and silver in theAmericas." In
Science of the Total Environment, Vol. I49 (1994), 168–169.

O'Phelan Godoy, Scarlett. "El norte y las revueltas anticlericales del siglo XVIII." In *Historia y Cultura*, No. 12. (Lima: 1979): 119–135.

O'Phelan Godoy, Scarlett. *Rebellions and Revolts in Eighteenth Century Peru and Upper
Peru*. Colonge: Bohlau Verlag, 1985.

Ossio, Juan M. "Cosmologies." In *International Social Science Journal*. Vol. IV9, No. 4
(1997): 549–562.

Palacio Atard, Vicente. "El asiento de la mina de Huancavelica en 1779." In *Revista de
Indias*. Vol. V (1944): 611–630.

Patiño Paúl Ortíz, Mariano. *Huancavelica colonial: Apuntes históricos de la ciudad
minera más importante del Virreynato Peruano*. Lima: Huancavelica 21, 2001.

Pearce, Adrian. "Huancavelica, 1563–1824: History and Historiography." In *Colonial Latin American Review*, Vol. II2, No. 3 (December, 2013): 422–440.

Pearce, Adrian. "Huancavelica 1700–1759: Administrative Reform of the Mercury Industry in Early Bourbon Peru." In *Hispanic American Historical* Review. Vol. VII9, No. 4
(Nov. 1999): 669–702.

Pearce, Adrian and Paul Heggarty. "'Mining the data' on the Huancayo-Huancavelica
linguistic frontier." In *History and Language in the Andes*. Paul Heggarty and Adrian
J. Pearce, Eds. New York: Palgrave MacMillan, 2011.

Pease, Franklin, and Faura N. Domínguez. *Los Incas en la colonia: Estudios sobre los
siglos XVI, XVII Y XVIIII en los Andes*. Lima: Museo National de Arqueología, Antropología e Historia del Péru, 2012.

Peet, Richard, and Michael Watts. "Liberating Political Ecology." In *Liberation Ecologies:
Environment, Development, Social Movements*. Peet, Richard, and Michael Watts,
Eds. Second edition. London: Routledge, 2004, 3–47.

Perreault, Thomas. "Conflictos del gas y su gobernaza: el caso de los guaranties de Tarija, Bolivia."*Anthropológica*. Vol. II8, No. 28. Suplemento 1(2010): 139–162.

Petersen, Georg. *Mining and Metallurgy in Ancient Perú*. William E. Brooks, Trans. Special Paper 467. Boulder, CO: The Geological Society of America, 2010.

Pieper, Renate. "Innovaciones tecnológicas y problemas del medio ambiente en la
minería novohispana (siglos XVI al XVIII)." In *Europa e Iberoamerica: Cinco siglos
de Intercambios*, Vol. II. María Justina Sarabia Viejo, et al., Eds. Seville, Spain: Asociación de Historiadores Latinoamericanistas Europeos/ Consejeria de Cultura y
Medio Ambiente, 1992, 353–368.

Piikivi, Leena, et al. "Psychological performance and long-term exposure to mercury
vapors." In *Scandanavian Journal of Work and Environmental Health*, Vol. Io (1984):
35–41.

Ping, Li, Xinbin Feng, et al. "Mercury exposures and symptoms in smelting workers of artisanal mercury mines in Wuchuan, Guizhou, China." In *Environmental Research*. Vol. I07 (2008): 108–114.

Ping, Li, Xinbin Feng, et al. "Mercury exposure in the population from Wuchan mercury mining area, Guizhou, China." In *Science of the Total Environment*. Vol. III95 (2008): 72–79.

Pinto Huaracha, Miguel, and Alejando Sánchez Salinas. *Las Rutas Del Café Y El Trigo: Los Ferrocarriles De Chanchamayo Y Huancavelica 1886–1932*. Lima: Seminario de Historia Rural Andina, Universidad Nacional Mayor de San Marcos, 2009.

Plasencia Soto, Rommel, and Fernando Cáceres Ríos. *Bibliografía de Huancavelica*. Lima: Universidad Nacional Mayor de San Marcos, 1996.

Povea Moreno, Isabel M. "Entre la retórica y la disuación. Defensores e impugnadores del sistema mitayo en Huancavelica y en las Cortés de Cádiz." *La Constitución Gaditana de 1812 y sus repercusiones en América*. Alberto Gullón Abao and Antonio Gutiérrez Escudero, Eds. Cadiz: Universidad de Cadiz, 2012, 201–211.

Povea Moreno, Isabel M. "Los buscones de metal. El sistema de pallaqueo en Huancavelica (1793–1820)." In *Anuario de Estudios Americanos*, Vol. VI9, No.1 (2012): 109–138.

Povea Moreno, Isabel M. *Minería y reformismo borbónico en el Perú. Estado, empresa y trabajadores en Huancavelica 1784–1814*. Lima: Instituto de Estudios Peruanos, 2014.

Povea Moreno, Isabel M. *Retrato de una decadencia: régimen laboral y sistema de explotación en Huancavelica, 1784–1814*. Diss. Granada: University of Granada, 2012.

Probert, Alan. "Bartolomé de Medina: The Patio Process and the Sixteenth Century Silver Crisis." In *Mines of Silver and Gold in the Americas*. Peter Bakewell, Ed. Brookfield, VT: Variorum, 1997.

Puche, Octavio. "Influencia de la legislación minera, del laboreo, así como del desarrollo técnico y ecnonómico, en el estado y producción de las minas de Huancavelica, durante sus primeros tiempos." In *Minería y metalurgia: Intercambio tecnológico y cultural entre América y Europa durante el período colonial español*. Manuel Castillo Martos, Ed. Seville: Muñoz Moya y Montraveta, Editores, 1994, 437–482.

Purser, W.C.F., *Metal Mining in Peru, Past and Present*. New York: Praeger Publishers, 1971.

Reyes Flores, Alejandro. "Huancavelica, 'Alhaja de la Corona': 1740–1790." In *Ensayos en ciencias sociales*. Julio Mejía Navarrete, Ed. Lima: Universidad Nacional Mayor de San Marcos, 2004.

Rivas Berrocal, Oswaldo. *Huancavelica: Bases Para El Desarrollo Económico Y Social Del Departamento Huancavelica*. Lima: Editorial Monterrico, 1989.

Robins, Nicholas. Field Research, August, 2010, July, 2012, June–July, 2015.

Robins, Nicholas. "La leyenda negra: esclavos negros en las minas de Potosí." In *Mitos Expuestos: Leyendas Falsas de Bolivia*. Nicholas Robins and Rosario Barahona, Eds. (Cochabamba, Bolivia: Editorial Kipus, 2014).

Robins, Nicholas. *Mercury, Mining and Empire: The Human and Ecological Cost of Colonial Silver Mining in the Andes*. Bloomington: Indiana University Press, 2011.

Robins, Nicholas. *Priest-Indian Conflict in Upper Peru: The Generation of Rebellion, 1750–1780*. Syracuse: Syracuse University Press, 2007.

Robins, Nicholas, N. Hagan, S. Halabi, H. Hsu-Kim, R.D. Espinoza Gonzales, M. Morris, G. Woodall, D. Richter, P. Heine, T. Zhang, A. Bacon, and J. Vandenberg. "Estimations of Historical Atmospheric Mercury Concentrations from Mercury Refining and Present-Day Soil Concentrations of Total Mercury in Huancavelica Peru." In *Science of the Total Environment*. Vol. IV26, No. 11 (June, 2012):146–154.

Rodríguez-Rivas, Daniel Alonso. "La legislación minera hispano-colonial y la intrusion de labores." In *La minería hispana e iberoamericana. Ponencias del 1 coloquio internacional sobre historia de minería*. Vol. I. N.A. Leon, Spain: Catedra de San Isidoro, 1970, 657–668.

Roel Pineda, Virgilio. *Historia social y económica de la colonia*. Lima: Gráfica Labor, 1970.

Rowland, Andrew, et al. "The effect of occupational exposure to mercury vapour on the fertility of female dental assistants." In *Occupational and Environmental Medicine*. Vol. V1, No. 1 (January 1994): 28–34.

Rudolph, William E. "The Lakes of Potosi." In *The Geographical Review*, Vol. II6, No. 4 (October, 1936): 529–554.

Ruiz Estrada, Arturo. *Arqueología de la ciudad de Huancavelica*. Lima: Servicios de Artes Gráficas, 1977.

Saignes, Thierry, "Capoche, Potosí y la coca: El consumo popular de estimulantes en el siglo XVII." In *Revista de Indias*, Vol. IV8, Nos. 182–183 (1988): 207–236.

Saignes, Thierry. "Las etnias de Charcas frente al sistema colonial (siglo XVI). Ausentismo y fugas en el debate sobre la mano de obra indígena (1595–1665)." In *Jahrbuch fur Geschichte von Staat, Wirtschaft und Gesellschaft Lateinamerikas*, Vol. II1 (1984), 27–75.

Sala Catala, José. "Vida y muerte en la mina de Huancavelica durante la primera mitad del siglo XVIII." In *Asclepio*. Vol. III9 (1987): 193–204.

Salas Guevara, Federico. *Historia De Huancavelica*. 2 Vols. Lima: Compañía de Minas Buenaventura, 2008.

Salas Guevara, Federico. *Villa rica de Oropesa*. Lima, Peru: NP, 1993.

Salgueiro, Barboni, Mirella Telles, Marcelo Fernandes da Costa, et al. "Visual field losses in workers exposed to mercury vapor." In *Environmental Research*, Vol. 107 (2008): 124–131.

Saguier, Eduardo. "Los calculos de rentabilidad en la crisis de la azogueria potosina. El refinado del metales a la luz de ocho visitas de ingenios desconocidas." In *Andes: Antropologia e historia.* Nos. 2–3, (1990–1991): 117–172.

Sallnow, Michael J. *Pilgrims of the Andes: Regional Cults in Cusco.* Washington: Smithsonian Institution Press, 1987.

Salvia, Daniela di. "La Pachamama En La Época Incaica y Post-Incaica: Una Visión Andina a Partir De Las Crónicas Peruanas Coloniales (Siglos XVI y XVII)/ Pachamama in the Inca and Post-Inca Period: An Andean Vision from the Colonial Chronicles of Peru (16th and 17th Centuries)." In *Revista Española de Antropología Americana* Vol. IV3, No.1 (2013): 89–110.

Sánchez Albornoz, Nicolás. *Indios y tributos en el Alto Perú.* Lima: Instituto de Estudios Peruanos, 1978.

Sánchez Albornoz, Nicolás. "Mita, migraciones y pueblos: variaciones en el espacio y en el tiempo Alto Perú, 5173–1692." In *Historia Boliviana,* Vol III, No. 1, (Cochabamba, 1983): 31–59.

Sánchez Albornoz, Nicolás. *The Population of Latin America: A History.* Translated by W.A.R. Richardson. Berkeley: University of California Press, 1974.

Sánchez Gómez, Julio. "La ténica en la producción de metales monedables en España y en América, 1500–1650." In *La savia del imperio: tres estudios de economia colonial.* Julio Sánchez Gómez, Guillermo Mira Delli-Zotti and Rafael Dobado, Eds. Salamanca: Ediciones Universidad Salamanca, 1997, 17–264.

Schutte, Norbert, et al. "Mercury and its Compounds." In *Occupational Medicine* 3rd. ed, Carl Zenz, et al. Eds. St. Louis, MO: Mosby, 1994, 549–557.

Scribner, Craig D. *To Air Is Human: Oppressive Conditions in Huancavelica's Mercury Mine from 1600–1616 and Efforts to Remedy Them.* Honors Thesis, Brigham Young University, 1995.

Seclén Santisteban, Segundo. "Pobreza e inequidad en salud." In *Coyuntura: Análisis Económico y Social de* Actualidad. Vol. IV, No. 21 (2013): 9–12.

Serrudo, Eberth. "El tampu real de Inkahuasi y la ocupación Inka en Huaytará." In *Inka Llaqta.* Vol. I (2010): 173–193.

Serulnikov, Sergio, "Customs and Rules: Bourbon Rationalizing Projects and Social Conflicts in Northern Potosí During the 1770s." In *Colonial Latin American Review.* Volume 8, No. 2, 1999: 245–274.

Singewald, Joseph T. "The Huancavelica Mecury Deposits, Peru." In *Engineering and Mining Journal.* Vol, 110, No. 11 (1920): 518–522.

Smith, C.T. "Depopulation of the Central Andes in the 16th Century." In *Current Anthropology.* Vol. I1, Nos. 4–5 (1970): 453–464.

Smith, P.J. et al. "Effects of occupational exposure to elemental mercury on short term memory." In British Journal of Industrial Medicine, Vol. IVo (1983): 413–419.

Soleo, L, et al. "Effects of low exposure to inorganic mercury on psychological performance." In *British Journal of Industrial Medicine*, Vol. IV7, No. 2 (1990): 105–109.

Spalding, Karen. *Huarochirí*. Stanford: Stanford University Press, 1984.

Stavig, Ward "'Living in Offense of Our Lord:' Indigenous Sexual Values and Marital Life in the Colonial Crucible." In *Hispanic American Historical Review*. Vol. VII5, No. 4 (November, 1995): 597–622.

Stein, Stanley and Barbara, *Silver, Trade and War: Spain and America in the Making of Early Modern Europe*. Baltimore: Johns Hopkins University Press, 2000.

Tamayo, Augusto. "Mina de cinabrio 'Santa Bárbara' en Huancavelica." In *Boletín del Ministerio de Fomento*, Vol. II, No. 1 (Lima, 1904): 38–43.

Tandeter, Enrique. *Coercion and Market: Silver Mining in Colonial Potosí, 1692–1826*. Albuquerque, University of New Mexico Press, 1993.

Tandeter, Enrique. "Crisis in Upper Peru, 1800–1805." In *Hispanic American Historical Review*, Vol. VII1, No. 1 (1991): 35–71.

Tandeter, Enrique. "Forced and Free Labor in Late Colonial Potosí." In *Past and Present*. Vol. IX3 (1981): 98–136.

Tandeter, Enrique. "Mineros de Week-end: Los ladrones de minas de Potosí." In *Todo es Historia*. Vol. 174 (Buenos Aires, 1978): 32–45.

TePaske, John J., and Kendall W. Brown, Eds. *Atlantic World, Volume 21: New World of Gold and Silver*. Boston, MA, USA: Brill Academic Publishers, 2010.

Thoms, Bryn and Nicholas Robins. "Remedial Investigation. Huancavelica Mercury Remediation Project. Huancavelica, Peru." Unpublished Manuscript. July, 2015.

Thoms, Bryn, Nicholas Robins, Enrique Ecos, Earl W. Brooks and Rubén Darío Espinoza Gonzales. *Results of June–July, 2015 Field Study, Huancavelica, Peru*. Environmental Health Council, unpublished manuscript, 2015.

Thoms, Bryn, Nicholas Robins, Enrique Ecos, and William E. Brooks. *Results of June/July 2016 Assessment of Soil and Fish, Huancavelica Mercury Remediation Project*. Environmental Health Council, unpublished manuscript, 2016.

Thomson, Sinclair. *We Alone Will Rule: Native Andean Politics in the Age of Insurgency*. Madison: University of Wisconsin Press, 2002.

Torrico y Mesa, Juan. "Excursión al departamento de Huancavelica." In *Boletín de minas, industrias y construcciones*. Vol. V, No. 12 (Lima: 1889): 889–893.

Tudela y Bueso, Juan Perez. "El problema moral en el trabajo minero del indio (siglos XVI y XVII)." In *La minería hispana e iberoamericana*. Vol. I. N.A. Leon, Spain: Catedra de San Isidoro, 1970, 355–371.

Umlauff, Augusto F. "El Cinabrio De Huancavelica." In *Boletin del Cuerpo de Ingenieros de Minas del Peru*. No. 7. Lima: Librería Escolar é Imprenta de E. Moreno, 1904.

United States Environmental Protection Agency. *A Review of the Reference Dose and Reference Concentration Processes*. EPA/630/P-02/002 F. Risk Assessment Forum. Washington, DC.: EPA, 2002.

United States Environmental Protection Agency. "Arsenic, inorganic." Integrated Risk Assessment System (IRIS), Chemical Assessment Summary. Washington, D.C.: National Center for Environmental Assessment, 1988. http://cfpub.epa.gov/ncea/iris/iris_documents/documents/subst/0278_summary.pdf.

United States Environmental Protection Agency. "Lead and compounds." Integrated Risk Assessment System (IRIS), Chemical Assessment Summary. Washington, D.C.: National Center for Environmental Assessment, 2004. http://cfpub.epa.gov/ncea/iris/iris_documents/documents/subst/0277_summary.pdf.

United States Environmental Protection Agency. *Mercury Study Report to Congress*. Vol. I. Executive Summary. Washington, D.C.:United States Environmental Protection Agency, 1997.

Uzzell, Barbara and Jacqueline Oler. "Chronic Low-Level Mercury Exposure and Neuropsychological Functioning." In *Journal of Clinical and Experimental Neuropsychology*, Vol. VIII, No. 5 (1986): 581–593.

Valcárcel, Luis. "El 'Memorial' del Padre Salinas." In *Memorial de las Historias del nuevo mundo Piru*. Fray Buenaventura Salinas y Córdoba. Lima: Universidad Nacional Mayor de San Marcos, 1957, Ix–xxvii.

Vargas Ugarte, S.J., Rubén. *Historia general del Perú. Virreinato (1551–1596)*. Vol II. Lima: Carlos Milla Batres, 1966.

Velázquez, Tesania. *Salud Mental En El Perú: Dolor Y Propuesta: La Experiencia de Huancavelica*. Lima: Consorcio de Investigación Económica y Social, 2007.

Waite, Benjamin. "Puya raimondii: Wonder of the Bolivian Andes." In *Journal of the Bromiliad Society*. Vol. II8, No. 5 (September–October, 1978) 200–208.

Waldron, H.A., and A. Scott. "Metals." In *Hunter's Diseases of Occupations*, 8th Ed. P.A.B. Raffle et al., Eds. London: E. Arnold, 1994, 90–138.

Walker, Charles. *Shaky Colonialism: The 1746 Earthquake-Tsunami in Lima, Peru and its Long Aftermath*. Durham, Duke University Press, 2008.

West, Irma and James Lim. "Mercury Poisoning Among Workers in California's Mercury Mills." In *Journal of Occupational Medicine*. Vol. I0, No. 12 (December, 1968): 697–701.

West, Robert. "Aboriginal Metallurgy and Metalworking in Spanish America: A Brief Overview." In *Mines of Silver and Gold in the Americas*. Peter Bakewell, Ed. Brookfield, VT: Variorum, 1997.

Whitaker, Arthur. *The Huancavelica Mercury Mine*. Boston: Harvard University Press, 1941.

Wiedner, Donald L. "Forced Labor in Colonial Peru." In *The Americas*, Vol.16, No. 4, (April, 1960): 357–383.

Wightman, Ann. *Indigenous Migration and Social Change: The Foresteros of Cuzco, 1570–1720*. Durham: Duke University Press, 1990.

Williamson, A.M., et al. "Occupational Mercury Exposure and its Consequences for Behaviour." In *International Archives of Occupational and Environmental Health*. Vol. Vo (1982): 273–286.

Wise, James M., and Jean Féraud. "Historic maps used in new geological and engineering evaluation of the Santa Bárbara mine, Huancavelica mercury district, Peru." In *De Re Metallica*, Vol.4 (2005): 15–24.

Yates, Robert G, Dean F. Kent, and Concha J. Fernández. *Geology of the Huancavelica Quicksilver District, Peru*. Washington, D.C.: U.S. G.P.O, 1951.

Yeates, Keith, and Mary Ellen Mortensen. "Acute and Chronic Neuropsychological Consequences of Mercury Vapor Poisoning in Two Early Adolescents." In *Journal of Clinical and Experimental Neuropsychology*. Vol. I6, No.2 (1994): 209–222.

Zavala, Silvio. *El servicio personal de los indios en el Perú*. Vol. I, Mexico City: El Colegio de Mexico, 1978.

Zimmerman, Authur F. *Francisco de Toledo, Fifth Viceroy of Peru, 1569–81*. Caldwell, Idaho: The Caxton Printers, Ltd., 1938.

Zulawski, Ann. *They Eat from Their Labor: Work and Social Change in Colonial Bolivia*. Pittsburgh: Pittsburgh University Press, 1995.

Zulawski, Ann. "Wages, Ore Sharing and Peasant Agriculture: Labor in Oruro's Silver Mines, 1607–1720." In *Hispanic American Historical Review*. Vol. VI7, No. 3, (1987): 405–430.

Websites

http://www.cdc.gov/measles/about/transmission.html
http://www.cdc.gov/flu/about/disease/spread.htm
http://emergency.cdc.gov/agent/smallpox/overview/disease-facts.asp
http://www.cdc.gov/measles/about/complications.html
http://www.cdc.gov/flu/about/disease/spread.htm
http://www.cdc.gov/flu/about/disease/symptoms.htm
http://www.cdc.gov/measles/about/signs-symptoms.html
http://www.cdc.gov/plague/resources/235098_Plaguefactsheet_508.pdf
http://www.cdc.gov/plague/symptoms/index.html
http://www.cdc.gov/plague/transmission/index.html
http://whc.unesco.org/en/criteria/
http://whc.unesco.org/en/list/1313
http://www.cdc.gov/travel/yellowbook/2014/chapter-3-infectious-diseases-related-to -travel/rickettsial-spotted-and-typhus-fevers-and-related-infections-anaplasmosis -and-ehrlichiosis
http://www.nlm.nih.gov/medlineplus/ency/article/000596.htm
http://www.nlm.nih.gov/medlineplus/ency/article/001363.htm

http://www.atsdr.cdc.gov/toxfaqs/index.asp
http://www.who.int/mediacentre/factsheets/fs379/en/
http://water.epa.gov/lawsregs/rulesregs/sdwa/lcr/lcrmr_index.cfm
http://water.epa.gov/lawsregs/rulesregs/sdwa/arsenic/regulations.cfm
http://www.who.int/water_sanitation_health/dwq/arsenicsum.pdf

Appendix

TABLE 1 _Total mercury (Hg) in soil in mg/kg remedial investigation Huancavelica, Peru_

Sample ID	Sample date	Total Hg (mg/kg)				Location	Observations/Comments
		Surface	1" Depth	3" Depth	Ave		
H1-fn-1a	06/13/2009	3.6	5.5	3.3	4.4	Transect 1 — Far North (4)	Colonial oven
H1-nn-1a	06/13/2009	238.4	248.8	88.1	168.4	Near North (5)	Foundation
H1-c-1a	06/13/2009	12.6	10.5	0.0	10.5	Center (3)	Foundation former house empty lot St. Cristóbal neighborhood
H1-ns-1a	06/13/2009	186.9	106.1	90.2	98.1	Near South (2)	Corner lot next to street, 20 m from river
H1-fs-1a	06/13/2009	15.0	23.5	32.5	28.0	Far South (1)	Near terminal, some runoff of tank of unknown origin
H2-fn-1a	06/13/2009	2.9	2.0	1.5	1.7	Transect 2 — Far North (6)	Hill overlooking city Ascención Plaza
H2-nn-1a	06/13/2009	1200.7	697.6	0.0	697.6	Near North (7)	Riverside near prison
H2-c-1a	06/13/2009	146.9	174.7	0.0	174.7	Center (8)	Empty lot on street
H2-ns-1a	06/13/2009	661.6	688.5	0.0	688.5	Near South (9)	Empty lot on street
H2-fs-1a	06/14/2009	0.1	4.7	17.2	12.2	Far South (10)	Cliffside by the road
H3-fn-1a	06/13/2009	354.7	348.3	–	–	Transect 3 — Far North (11)	Road bed
H3-nn-1a	06/13/2009	76.2	72.6	–	–	Near North (12)	Near Ichu River by school
H3-c-1a	06/13/2009	15.0	16.9	–	–	Center (13)	Field next to convent
H3-ns-1a	06/13/2009	2.1	2.5	–	–	Near South (14)	Hill overlooking town and cemetery
H3-fs-1a	06/13/2009	39.1	34.4	–	–	Far South (15)	Near Coliseum
H1-OH-1	06/14/2009	284.3	–	–	–	Transect 4	San Cristóbal sample from home foundation – 1 foot
H1-OH-2	06/14/2009	47.4	–	–	–	–	Sample from home 3 feet above
H1-OH-3	06/14/2009	244.1	–	–	–	–	Exterior wall of home 2 feet

© KONINKLIJKE BRILL NV, LEIDEN, 2017 | DOI 10.1163/9789004343795_010

TABLE 1 *Total mercury (Hg) in soil in mg/kg remedial investigation Huancavelica, Peru* (cont.)

| Sample ID | Sample date | Total Hg (mg/kg) | | | | Location | Observations/Comments |
		Surface	1" Depth	3" Depth	Ave		
H-CO-1	06/14/2009	597.7	--	--	--	--	Colonial oven
H-TO-1	06/14/2009	121.6	--	--	--	--	Colonial oven
H-M1-1	06/14/2009	517.0	--	--	--	--	Tailings central courtyard in house

Action Levels

USEPA RSL for Residential	23 mg/kg
USEPA RSL for Industrial	310 mg/kg
Site-specific RBC for residential	75 mg/kg
Site-specific RBC for residential adult	520 mg/kg
Site-specific RBC for occupational	770 mg/kg

Notes

mg/kg – milligrams per kilogram or parts per million.

Average of 2009 samples is average of both 1" and 3" results.

-- – not sampled.

USEPA – US Environmental Protection Agency.

RSL – Regional Screening Level, which uses a Hazard Quotient of 1 for a 30 year (chronic) exposure.

RBC – Risk-Based Concentration developed using the generic ATSDR/ODEQ exposure calculation.

Site-specific RBC assumes 10% of the mercury is in readily bioaccessibile forms.

USEPA's Industrial exposure scenario uses the same exposure factors as ODEQ's Occupational exposure scenario.

Cells shaded indicate results above site-specific RBC for residential use.

TABLE 2   *Total mercury in Earthen homes (dust, walls, floors, and vapor) remedial*
             investigation Huancavelica, Peru

Blind ID 2	Neighborhood	Dust (mg/kg)	Wall (mg/kg)	Floor (mg/kg)	Vapor(μg/m^3)
RI H1	Ascención	78.5	41.3	93.3	0.56
RI H2	Ascención	17.3	76.9	47.2	0.00
RI H3	Ascención	64.7	140.5	149.7	0.00
RI H4	Ascención	71.1	9.9	298.0	1.86
RI H5	Yananaco	256.2	1071.7	926.3	0.00
RI H6	Yananaco	124.1	98.6	455.3	0.00
RI H7	Yananaco	52.7	85.1	108.2	0.36
RI H8	Yananaco	144.3	714.5	364.3	0.19
RI H9	Ascención	24.6	270.3	112.3	0.58
RI H10	Ascención	53.3	122.0	190.4	0.00
RI H11	Ascención	9.7	107.8	125.9	0.00
RI H12	Ascención	81.4	763.3	204.2	1.35
RI H13	Ascención	35.0	259.2	833.3	0.00
RI H14	Ascención	413.1	49.9	178.2	5.08
RI H15	Ascención	27.7	54.2	60.1	0.44
RI H16	Ascención	12.9	10.4	16.0	0.00
RI H17	Ascención	20.9	85.4	31.9	1.44
RI H18	Ascención	25.8	391.3	262.3	0.68
RI H19	Ascención	8.1	13.7	18.2	0.80
RI H20	Yananaco	119.3	85.6	68.8	0.00
RI H21	Yananaco	34.7	96.2	280.7	0.00
RI H22	Yananaco	59.5	26.3	48.8	0.00
RI H23	Yananaco	53.2	213.7	86.7	0.00
RI H24	Yananaco	13.2	296.4	40.1	0.00
RI H25	Yananaco	43.3	39.8	40.9	0.00
RI H26	Yananaco	58.2	139.8	125.0	0.00
RI H27	Yananaco	39.1	55.3	27.5	0.00
RI H28	Yananaco	12.2	30.3	19.2	0.00
RI H29	Yananaco	133.5	731.7	412.0	0.48
RI H30	Yananaco	137.6	931.7	545.0	0.00
RI H31	Santa Ana	4.2	8.0	40.5	0.00
RI H32	Santa Ana	4.8	9.7	22.2	0.00
RI H33	Santa Ana	5.8	10.9	4.7	1.31
RI H34	Santa Ana	12.1	15.9	7.0	0.78
RI H35	Santa Ana	4.2	28.2	34.3	0.00

TABLE 2   *Total mercury in Earthen homes (dust, walls, floors, and vapor) remedial investigation Huancavelica, Peru (cont.)*

Blind ID 2	Neighborhood	Dust (mg/kg)	Wall (mg/kg)	Floor (mg/kg)	Vapor (μg/m³)
RI H36	Santa Ana	21.1	17.8	43.7	1.26
RI H37	Santa Ana	2.4	14.2	3.2	0.00
RI H38	Santa Ana	3.3	11.3	7.0	2.72
RI H39	Santa Ana	11.8	11.9	3.1	1.04
RI H40	Santa Ana	3.9	8.4	4.6	0.00
RI H41	Santa Ana	6.4	243.2	66.4	0.00
RI H42	Santa Ana	11.6	13.6	22.9	0.00
RI H43	Santa Ana	3.1	18.3	20.4	0.00
RI H44	Santa Ana	19.5	16.5	54.2	0.26
RI H45	Santa Ana	2.5	39.4	6.3	0.00
RI H46	San Cristóbal	51.6	284.0	186.3	0.87
RI H47	San Cristóbal	83.5	253.3	193.4	1.08
RI H48	San Cristóbal	152.7	281.7	283.7	1.78
RI H49	San Cristóbal	100.7	943.7	238.7	0.00
RI H50	San Cristóbal	109.2	546.0	839.0	0.00
RI H51	San Cristóbal	9.6	338.2	25.4	0.00
RI H52	San Cristóbal	17.5	27.0	33.5	0.63
RI H53	San Cristóbal	59.9	209.3	144.3	0.00
RI H54	San Cristóbal	73.4	199.7	146.7	1.65
RI H55	San Cristóbal	61.1	105.5	157.0	0.00
RI H56	San Cristóbal	46.5	247.0	112.1	0.00
RI H57	San Cristóbal	149.2	479.3	16.7	0.25
RI H58	San Cristóbal	24.9	175.7	53.6	1.62
RI H59	San Cristóbal	14.5	50.3	39.2	0.00
RI H60	San Cristóbal	25.7	21.6	25.2	0.00

Action Levels for Soil

USEPA RSL for Residential	23 mg/kg
USEPA RSL for Industrial	310 mg/kg
Site-specific RBC for residential child	75 mg/kg
Site-specific RBC for residential adult	520 mg/kg
Site-specific RBC for occupational	770 mg/kg
Action Levels for Vapor (residential only, all age classes)	
USEPA RfC	0.3 μg/m³
The WHO Chronic Screening Level	0.2 μg/m³
USEPA Action Level for Relocation of Resident	1.0 μg/m³

Notes

mg/kg – milligrams per kilogram or parts per million.

mg/m³ – micrograms per meter cubed or parts per billion.

USEPA – US Environmental Protection Agency.

RSL – Regional Screening Level, which uses a Hazard Quotient of 1 for a 30 year (chronic) exposure.

RBC – Risk-Based Concentration developed using the generic ATSDR/ODEQ exposure calculation.

RfC – Reference Concentration for chronic exposure.

WHO – World Health Organization.

MRL – Minimum Risk Level for chronic exposure.

USEPA's Industrial exposure scenario uses the same exposure factors as ODEQ's Occupational exposure scenario.

Site-specific RBC assumes 10% of the mercury is in readily bioaccessibile forms.

Cells shaded indicate results above site-specific RBC for soil for residential use or the RfC for chronic exposure to vapor (The WHO value of 0.2 µg/m3 is used).

House IDs shaded in pink (color) and grey (B&W) have all four media (dust, wall, floor, and vapor) above residential (site-specific) action levels (RBC or MRL).

Analytical method – Atomic absorption spectrometry.

Data from Hagan N, Robins N, Hsu-Kim H, Halabi S, Espinoza Gonzales RD, et al. (2013) "Residential Mercury Contamination in Adobe Brick Homes in Huancavelica, Peru." *PLoS ONE* Vol. 8 No. 9. (September, 2013) doi:10.1371/journal.pone.0075179.

TABLE 3 *Mercury, Arsenic and Lead in Huancavelica and area rock samples 2015 pilot study Huancavelica, Peru*

Sample ID	Sample description	Land use*	Heavy metals of concern (mg/kg)				Location
			As	Hg	Pb		
HV1	Travertine block in wall	Residential	530	4.1	32		
HV2	Breccia block within exterior wall	Residential	12	1.3	7		
HV3	Soil in floor of east room of home	Residential	100	1.0	29		
HV4	Travertine block	Residential	887	0.7	12		
HV5	Calcined material looks like scoria	Residential	454	206.0	5510		
HV6	Outcrop with sulfides veinlets	Residential	8	1.3	24		
HV8	Red wall material near	Residential	1060	246.0	2160		
HV9	Travertine outcrop near Petro Peru Espinoza	Occupational	839	1.8	15		east end of town
HV10	Waste near colonial furnace in San Cristobal	Occupational	562	44.5	179300		San Cristóbal
HV11	Tailings at plaza of Sacsamarca	Occupational	436	16.5	1070		Sacsamarca
QA Samples							
HV8-X	duplicate of HV-8		1070	236.0	2150		
HV15 EXTRA-X	duplicate of HV-15 EXTRA		16	4.5	103		

Screening Values (includes carcinogenic and noncarcinogenic, whichever is lower)

USEPA Regional Screening Level (RSL)	Residential	0.68	23	400
USEPA Regional Screening Level (RSL)	Occupational	3	350	800
Oregon DEQ Risk Based Concentration (RBC)	Residential	0.39	23	400
Oregon DEQ Risk Based Concentration (RBC)	Occupational	1.7	310	800
Site Specific Hg Screening Level from HVCA RI		--	75	--

Notes

Thanks to William Earl Brooks for permission to publish these data.

Heavy metals of Concern include As – arsenic, Ac – cadmium, Hg – mercury, and Pb – lead, additional metals analyses were conducted.

mg/kg – milligrams per kilogram or parts per million.

Land Use refers to the predominant assumed land use where the samples was collected, yet occupational could include either residential or recreational exposures.

Pink (color)/gray (B/W) shading for cadmium and lead represents concentrations over the lowest residential screening value presented at the bottom of the table.

Arsenic background is likely above the RSL or RBC and thus 100 mg/kg was arbitrarily used for shading elevated concentrations.

Cells shaded indicate results above the site specific screening value for HVCA presented at the bottom of the table.

Analytical Method – inductively-coupled plasma mass spectrometry at American Assay Laboratories.

TABLE 4   *Food stock, water, and sediment sample results 2016 pilot study Huancavelica, Peru*

Sample ID	Description	Date	Time	Total Hg	Comment
Food Stock Samples				*solids (mg/kg)*	
AT-1	Llama	6/30/15	1000	<0.03	From Tansiri area
MT-1	Mashua	7/2/15	1040	<0.03	Grown near Senor de Oropesa
BT-1	Barley	7/2/15	1040	<0.03	Grown near Senor de Oropesa
PT-1	Potato	7/2/15	1040	<0.03	Grown near Senor de Oropesa
LT-1	Lamb	7/4/15	858	<0.03	Raised near Cachimayo
AMT-1	Alpaca meat	7/4/15	900	<0.03	Purchased from city market
BMT-1	Beef meat	7/4/15	905	<0.03	From Huanaspampa
TT-1	Rainbow Trout fillet	7/1/15	800	0.038	Line-caught in Choclococha Lagoon
TT-2	Rainbow Trout fillet	7/1/15	800	<0.03	Line-caught in Choclococha Lagoon
TT-3	Rainbow Trout fillet	7/1/15	800	0.046	Line-caught in Choclococha Lagoon
TT-4	Rainbow Trout fillet	7/6/15	1600	<0.03	Acoria municipal fish farm
TT-5	Rainbow Trout fillet	7/6/15	1600	<0.03	Palca municipal fish farm
Water Samples				*liquids (ug/l)*	
AQUA	Water	07/20/2015	NA	<0.8	Huancavelica municipal tap water
AQUA	Water	07/20/2015	1550	<0.8	Huancavelica municipal tap water
AQUA	Water	07/20/2015	1510	<0.8	Spring water in home
AQUA	Water	07/20/2015	1630	<0.8	River water
AQUA	Water	07/20/2015	1503	<0.8	Montepata Reservoir tap water

Sediment Samples

				solids (mg/kg)	
IS-1	Ichu River Sediment	07/03/2015	1530	18.93	Downstream of Hvca
IS-2	Ichu River Sediment	07/03/2015	1610	0.88	Upstream of Hvca by quarry
IS-3	Ichu River Sediment	07/04/2015	1350	1370	Midtown by Ascención bridge
SD5	Ichu River Sediment	06/29/2016	113	14.7	Downstream of Hvca, similar location to IS-1

Screening Values

		solids (mg/kg)	
Food stock – USEPA ingestion advisory level, recreational		0.4	Based on recreational consumption rate, assumes all Hg is MeHg
Food stock – USEPA ingestion advisory level, subsistence		0.049	Based on subsistence consumption rate, assumes all Hg is MeHg
Water – USEPA MCL		0.5	Maximum contaminant level or safe drinking water level

Notes

mg/kg – milligrams per kilogram or parts per million.

ug/l – micrograms per liter or parts per billion.

< indicates the method detection limit.

NA – Not Available.

USEPA – US Environmental Protection Agency.

Analytical Method – Atomic absorption spectrometry performed at Cetox in Lima, Peru.

TABLE 5   *Residential vapor results 2015 pilot study Huancavelica, Peru*

	Average Vapor Result (ng/m3)			
Sample ID	Pre Remedy	Post Remedy	Background	% Reduction
H12	70	44	23	37
H16	267	55	34	79
H17	220	143	27	35
H18	434	124	42	71
H19	740	58	36	92
H20	62	138	12	-123
Screening value	USEPA RSL (residential) in ug/m3			310
Screening value	WHO Chronic in ng/m3			200

Notes

ng/m3 – nanograms per meter cubed or parts per trillion.

ND – not detected at or above the equipment detection limit, Jerome 431X detection limit is 500 ng/m3.

All results were collected using the Lumex RA 915+ for a 4 min analysis, except for the 2012 result which was collected by the Jerome 431x, see 2015 Pilot Study Report for detail.

USEPA – United States Environmental Protection Agency.

RSL – Regional Screening Level.

WHO – World Health Organization.

Post remedy increases have been attributed to residual elevated vapor from disturbance of wall and floors during construction activity.

TABLE 6 Soil sample results using X-ray florescence 2016 field event Huancavelica, Peru

Sample ID	Sample type	Sample date	Sample time	As (ppm)	Hg (ppm)	Pb (ppm)	Sample description
S1 (Ave of dups)	Surface soil	06/25/2016	1406	935	543	3003	Main Plaza Sacsamarca, 30 point composite
S2 (Ave of dups)	Surface soil	06/25/2016	1408	674	527	2330	Soccer goal area Sacsamarca plaza
S3	Surface soil	06/25/2016	1417	2268	220	6138	Tailings in Sacsamarca plaza NW corner
S3 (Ave of dups)	Surface soil	06/25/2016	1417	2261	225	6114	Tailings in Sacsamarca plaza NW corner
S4	Surface soil	06/25/2016	1429	39	21	47	Sacsamarca playground
S5	Surface soil	06/25/2016	1430	364	189	1840	Sacsamarca main street
S6	Surface soil	06/26/2016	1515	64	14	174	Soil in alluvial plain at SD5 location
S7	Background	06/27/2016	1329	5.8	<LOD	12	Soil near road on way to Acoria
S8	Background	06/27/2016	1401	13	<LOD	26	Soil near road on way to Acoria
S9	Background	06/27/2016	1420	9.3	7	19	Soil near road on way to Acoria
S10	Building material	06/27/2016	1528	1743	471	4721	Pile of red "tailings" near SD8
S11	Surface soil	06/27/2016	1556	75	3	287	At turnaround on way to Hvca
S12	Background	06/28/2016	1315	6.7	<LOD	14	Road to Hvca, removed grass cover
S13	Background	06/28/2016	1332	50	<LOD	29	
S14	Surface soil	06/29/2016	1500	122	28	172	Sample on path along river by bridge
S15 (Ave of dups)	Surface soil	06/29/2016	1516	124	168	384	Empty lot on corner of Jr Moira P De Bollido and Jr Jose Olaya

TABLE 6 *Soil sample results using X-ray florescence 2016 field event Huancavelica, Peru* (cont.)

Sample ID	Sample type	Sample date	Sample time	As (ppm)	Hg (ppm)	Pb (ppm)	Sample description
S16	Surface soil	06/29/2016	1528	350	10	255	Above hospital
S17	Building material	06/29/2016	1538	172	33	335	Possible building material by Santa Barbara Road
S18	Surface soil	06/29/2016	1549	99	57	400	–
S19	Building material	06/29/2016	1559	69	3	31	Possible building material pile
S20 (Ave of dups)	Building material	06/29/2016	1616	1082	308	3414	Possible building material, reddish
S21	Surface soil	06/29/2016	1619	519	39	513	Exposed soil by market, busy walking corner
S22	Building material	06/29/2016	1629	690	318	850	Possible building material pile
S23	Surface soil	06/29/2016	1639	390	16	400	Roadway exposed soil
S24	Soil at Depth	06/29/2016	1645	1295	324	5737	Exposed reddish soil at 1.5 ft bgs, appears to be tailings (Cetox Hg Result = 349 mg/kg)
S25	Soil at Depth	06/29/2016	1658	157	11	273	Exposed road material, excavated road bed at 1.5 ft bgs
S26	Surface soil	06/29/2016	1714	450	8	77	–
S26 (Ave of dups)	Surface soil	06/29/2016	1714	458	13	70	–
S27	Surface soil	06/29/2016	1727	807	59	168	–

S28	Surface soil	06/29/2016	1734	615	22	104	Exposed soil in side walk
S29 (Ave of dups)	Background	06/29/2016	1400	15	<LOD	35	Sample of ashflow tuff at Killamachoy
S30	Background	06/30/2016	1200	14.2	<LOD	19.4	In field on way to Huancayo

Screening Values

USEPA Regional Screening Value (RSL) for Residential Exposure	0.39		400
Site Specific Screening Level for Residential Exposure*	NC	75	NC
Regional background, average of background soil samples	16	<LOD	22

Notes

ppm – parts per million.

Ave of dups – Average of duplicates, see "all samples" table for individual sample results for duplicates in 2016 Field Event Report.

<LOD – Less than the limit of Detection, approximately 10 ppm for Hg.

NC – not calculated.

* Site Specific Screening value calculated using site-specific bioavailability and modified daily ingestion rate. See Huancavelica Mercury Cleanup Project Remedial Investigation, dated July 30 2015, EHC.

Cells shaded indicate a result that is greater than either the regional background concentration or an available screening value.

Analysis performed at Reflex in Lima on July 1, 2016 using a Reflex-brand portable X-Ray fluorescence analyzer.

TABLE 7   *Total mercury in fish tissue 2016 field event Huancavelica, Peru*

Fish Farm	Location	Type	Species	Number of samples	Sample date	Result
Palca	14 km north of Hvca	river	Rainbow Trout	11	06/26/2016	All ND (<0.02 mg/kg)
Pultocc	38 km southwest of Hvca	lake	Rainbow Trout	12	06/28/2016	All ND (<0.02 mg/kg)
Acoria	19 km northeast of Hvca	river	Rainbow Trout	12	06/27/2016	All ND (<0.02 mg/kg)
Sacsamarca	3 km southwest of Hvca	river	Rainbow Trout	12	12/14/2016	Pending

Notes

ND – Not detected at laboratory method reporting limit.

mg/kg – milligram per kilogram.

Analytical Method – Atomic absorption spectrometry performed at Cetox in Lima, Peru.

TABLE 8   *Residential sample results using X-ray fluorescence 2016 field event Huancavelica, Peru*

Sample ID	Sample type 2	Sample date	Sample time	As (ppm)	Hg (ppm)	Pb (ppm)	Sample description
H1	Wall	06/28/2016	1612	636	113	1046	outside wall
H2	Wall	06/28/2016	1604	965	195	1638	inside untreated corner and outside of wall
H3	Wall	06/28/2016	1600	1579	299	3855	Outside wall
H4	Wall	06/28/2016	1530	51	40	164	
H5	Wall	06/25/2016	1250	586	112	603	Kitchen wall
H6	Wall	06/25/2016	1240	1045	345	1792	Upstairs room
H7	Wall	06/25/2016	1236	640	182	945	Downstairs bedroom wall
H8 (Ave of dups)	Floor	06/25/2016	1250	392	138	589	Bedroom floor
H9	Wall	06/25/2016	1210	803	555	1001	Upstairs room wall
H10	Wall	06/25/2016	1203	796	926	2026	Downstairs room wall
H11	Floor	06/25/2016	1253	279	55	289	Kitchen floor
H12	Wall	06/27/2016	1200	371	141	299	Outside wall
H13	Floor	06/27/2016	1300	362	9	200	Floor
H14 (Ave of dups)	Wall	06/27/2016	1140	467	112	159	Outside wall of house near front door
H15	Wall	06/27/2016	1000	295	11	165	Wall

TABLE 8　　*Residential sample results using X-ray fluorescence 2016 field event Huancavelica. Peru (cont.)*

Sample ID	Sample type 2	Sample date	Sample time	As (ppm)	Hg (ppm)	Pb (ppm)	Sample description
H16	Wall	06/27/2016	1400	144	696	572	Outside wall on lower part
H17 (Ave of dups)	Wall	06/27/2016	1350	1611	462	5282	Outside wall of kitchen/bedroom
Screening Values							
USEPA Regional Screening Value (RSL) for Residential Exposure				0.39		400	
Site Specific Screening Level for Residential Exposure*				NC	75	NC	
Regional background, average of background soil samples				16	<LOD	22	

Notes

ppm – parts per million.

Ave of dups – Average of duplicates, see "all samples" table for individual sample results for duplicates in 2016 Field Event Report.

<LOD – Less than the limit of Detection, approximately 10 ppm for Hg.

NC – not calculated.

* Site Specific Screening value calculated using site-specific bioavailability and modified daily ingestion rate. See Huancavelica Mercury Cleanup Project Remedial Investigation, dated July 30 2015, EHC.

Cells shaded indicate a result that is greater than either the regional background concentration or an available screening value.

Analysis performed at Reflex in Lima on July 1, 2016 using a Reflex-brand portable X-ray fluorescence analyzer.

Index

Printed in the United States
By Bookmasters